René Marcel Schmogrow

Real-time Digital Signal Processing for Software-defined Optical Transmitters and Receivers

Karlsruhe Series in Photonics & Communications, Vol. 13
Edited by Profs. J. Leuthold, W. Freude and C. Koos

Karlsruhe Institute of Technology (KIT)
Institute of Photonics and Quantum Electronics (IPQ)
Germany

Real-time Digital Signal Processing for Software-defined Optical Transmitters and Receivers

by

René Marcel Schmogrow

KIT Scientific Publishing

Dissertation, Karlsruher Institut für Technologie (KIT)
Fakultät für Elektrotechnik und Informationstechnik, 2013

Impressum

KIT Scientific Publishing

Karlsruher Institut für Technologie (KIT)
KIT Scientific Publishing
Straße am Forum 2
D-76131 Karlsruhe

KIT Scientific Publishing is a registered trademark of Karlsruhe Institute of Technology. Reprint using the book cover is not allowed.

www.ksp.kit.edu

Print on Demand 2014

ISSN 1865-1100
ISBN 978-3-7315-0257-9

Real-time Digital Signal Processing for Software-defined Optical Transmitters and Receivers

Zur Erlangung des akademischen Grades eines

DOKTOR-INGENIEURS

der Fakultät für Elektrotechnik und Informationstechnik
des Karlsruher Instituts für Technologie

genehmigte

DISSERTATION

von

René Marcel Schmogrow, M.Sc.

aus

Berlin

Tag der mündlichen Prüfung: 07.05.2013
Hauptreferent: Prof. Dr. sc. nat. Jürg Leuthold
Korreferenten: Prof. Dr.-Ing. Dr. h.c. Wolfgang Freude
Prof. Dr.-Ing. Michael Hübner

Iucundi acti labores. (Nach getaner Arbeit ist gut ruhen.)
– Marcus Tullius Cicero (106 B.C. – 43 B.C.)

Table of Contents

List of Figures

Abstract (German)

Die vorliegende Arbeit befasst sich mit der echtzeitfähigen Signalprozessierung für programmierbare optische Sender und Empfänger. Im Rahmen der Arbeit wurden zum einen effektive Algorithmen entwickelt, die sich durch einen geringen Prozessierungsaufwand auszeichnen und somit für den Einsatz in echtzeitfähigen optischen Empfängern geeignet sind. So wurde zum Beispiel eine Methode entwickelt, mit der Signale, die auf zwei orthogonalen Polarisationen gesendet wurden, getrennt und ohne Übersprechen empfangen werden können. Dabei ist neben der geringen Rechenkomplexität die Vielseitigkeit des Algorithmus in Bezug auf Pulsform, Modulationsformat sowie die Unabhängigkeit des Algorithmus von der Taktrückgewinnung hervorzuheben.

Ein weiterer Algorithmus wurde entwickelt, mit dem bei geringem Rechenaufwand, die Trägerinformation (Frequenz und Phase) sowie die Taktung von sogenannten OFDM- und Nyquist-Signalen zurückgewonnen werden. Auch dieser Algortihmus eignet sich für einen Einsatz in echtzeitfähigen optischen Empfängern.

Ein echtzeitfähiger optischer Sender wurde unter der Verwendung von im Feld programmierbaren Logik-Gatter-Anordnungen (field programmable gate arrays, FPGA) aufgebaut. Dieser Software-gesteuerte Sender erzeugt zahlreiche Signale verschiedener Modulationsformate und Pulsformen in Echtzeit. Hierbei lag der Fokus auf der Generation von spektral effizienten Signalen wie OFDM- und Nyquist-Signalen bei höchsten Datenraten von über 100 Gbit/s. Mit Hilfe dieser speziellen Pulsformen können bestehende optische Netze besser genutzt werden. Als nächster Schritt wurde der Prozessierungsaufwand und die Signalqualität von mit FPGAs erzeugten OFDM und Nyquist Signalen verglichen.

Mit Hilfe von offline erzeugten Nyquist Signalen und einer Quadratur Amplituden Modulation mit 512 Symbolen (512QAM) konnte die bis dahin am höchsten erreichbare spektrale Effizienz, d.h., das Verhältnis von Datenrate zu der benötigten Bandbreite, übertroffen werden. In diesem Experiment wurden Signale mit eine Datenrate von 54 Gbit/s in einer optischen Bandbreite von lediglich 3 GHz übertragen.

Verschiedene Varianten der Nyquist-Signalerzeugung wurden experimentell untersucht. Hierzu formen optische, analog elektrische und digitale Filter die Pulse. In einer Anordnung mit drei optischen Trägern wurde hierbei der jeweils minimal tolerierbare Trägerabstand ermittelt welcher bei digital geformten Pulsen nahe dem theoretischen Minimum liegt.

Der sogenannte mittlere Fehlervektor (error vector magnitude, EVM) gibt Aufschluss über die Qualität von QAM-Signalen und findet weite Verbreitung im Bereich der kupferbasierten und Funkübertragung. Im Rahmen dieser Arbeit wurde die Eignung der EVM für optische Übertragungskanäle experimentell analysiert. Insbesondere die Schätzung der Bitfehlerwahrscheinlichkeit anhand gemessener EVM-Werte lag im Fokus der Untersuchungen.

Abschließend wurde ein optisches Zugangsnetz aufgebaut, an dem die Signalqualität und Effizienz von OFDM- und Nyquist-Signalen untersucht wurden.

Preface

This thesis deals with real-time digital signal processing for software-defined optical transmitters and receivers. In the course of this thesis, efficient algorithms have been developed that are adequate for real-time optical receivers due to their low computational complexity. So for instance a method has been found to receive and separate signals that were transmitted on two orthogonal polarizations. No crosstalk was observed. The algorithm requires low computational effort and is versatile with respect to modulation format and pulse-shape. As an additional advantage, it functions independent from the timing recovery.

A second algorithm has been developed to recover the carrier (frequency and phase) as well as the timing of so-called orthogonal frequency division multiplexed (OFDM) and Nyquist signals. Also this algorithm requires low computational effort and is therefore well suited for real-time optical receivers.

A real-time software-defined transmitter based on field programmable gate arrays (FPGA) has been implemented. This transmitter is capable of generating various signals of different modulation formats and pulse-shapes in real-time. Special emphasis was put on the generation of spectrally efficient signals such as OFDM or Nyquist signals at highest data rates beyond 100 Gbit/s. These signals allow an improved utilization of existing optical networks. In a next step, the processing requirements and signal quality of FPGA-processed OFDM and Nyquist signals has been investigated.

The highest achieved spectral efficiency, i.e., the ratio of data rate and bandwidth, has been outperformed using offline generated Nyquist signals in 512QAM format. In this experiment, data with a total rate of 54 Gbit/s were transmitted using an optical bandwidth of only 3 GHz.

Different variants of Nyquist pulse-shaping were experimentally investigated. To this end optical, analog electrical, and digital filters were employed to shape the pulses. The signal quality and potential spectral efficiency of these signals has been assessed using three optical carriers in a model experiment.

The so-called error vector magnitude (EVM) is a quality measure for QAM signals. The EVM metric is well-known and employed in the field of wireline and wireless communications. The applicability of the EVM measure for optical transmission links has been investigated in the course of this thesis. The bit error ratio (BER) estimation from measured EVM values was examined in particular.

Finally, an optical access network was set up and used to assess the signal quality and efficiency of OFDM and Nyquist signals in this scenario.

Achievements of the Present Work

In this thesis, real-time digital signal processing for software-defined optical transmitters and receivers is investigated. Special attention is given to the real-time generation of spectrally efficient pulse shapes and to the capability to process data with rates beyond 100 Gbit/s. Several simulations and experiments were performed to verify the developed algorithms.

In the following, we give a concise overview of the main achievements.

Polarization Demultiplexing: A computational simple demultiplexing algorithm has been proposed and experimentally demonstrated. With this algorithm, signals that are transmitted on two orthogonal polarizations can be effectively separated without crosstalk [J3]. The proposed method operates blindly, i.e., without the need for training symbols or pilot tones, and its functionality is independent from pulse-shape and modulation format. Furthermore, it does not require a timing recovery and is expected to be well suited for gigabit real-time optical receivers. A patent has been filed on this topic.

Carrier and Timing Recovery: Methods used to recover the carrier frequency and phase as well as the proper timing have been developed. The proposed recovery schemes can be used for orthogonal frequency division multiplexed (OFDM) signals and Nyquist signals [J2]. The computational effort is low which makes the proposed method suitable for high-speed real-time applications. The functionality has been experimentally verified with OFDM and Nyquist signals modulated with different formats.

Real-time Digital Signal Processing for SW-Defined Optical Transmitters: An FPGA platform design for a software-defined optical transmitter has been implemented to flexibly generate and process optical signals of various pulse-shapes and modulation formats in real-time. With the implemented transmitter, quadrature amplitude modulated signals with 64 states (64QAM) can be generated at symbol rates of up to 28 GBd leading to data rates of 168 Gbit/s [J25].

Real-time OFDM Signal Generation beyond 100 Gbit/s: A real-time OFDM transmitter has been implemented and demonstrated. This transmitter generates OFDM signals beyond 100 Gbit/s (101.5 Gbit/s on a single polarization) [J15][C27], a record number which still has not been bettered. In this design, a look-up table (LUT) based inverse discrete Fourier transform (IDFT) was implemented and optimized. Additional real-time OFDM transmission experiments were performed in collaboration with the University College London and the Carnegie Mellon University [J20][C19].

Real-time Nyquist Pulse-Shaping beyond 100 Gbit/s: Digital finite-duration impulse response (FIR) filters have been designed and implemented to generate sinc-shaped

Nyquist pulses in real-time [J1][J22][C1][C18][C30]. Data rates up to 126 Gbit/s (per polarization) were demonstrated [J1][C1]. The digital processing has been optimized by introducing a dynamic computational precision [J1][C18]. Furthermore, in order to optimally facilitate the limited sampling speed of digital-to-analog converters (DAC) and analog-to-digital converters (ADC), parallel processing techniques allowing rational oversampling factors were developed [J1][C1].

Comparison of Real-time Nyquist Pulse-Shaping and OFDM: FPGA-based implementations of real-time OFDM and Nyquist signaling were compared with respect to resource requirements and performance in an optical transmission link with a single and with three optical carriers [J16][C26]. This was the first time that actual real-time pulse shaping with FPGAs had been compared.

Nyquist Pulse-Shaping for Record Spectral Efficiencies: The spectral efficiency, i.e., the achieved data rate within a certain bandwidth, has been brought to record numbers by employing Nyquist sinc-pulse shaping and 512QAM formatting [J17][C12].

Comparison of Analog and Digital Filtering for Nyquist Signaling: The performance of Nyquist pulse-shaping with analog and digital filters was compared [J12]. Thereby Nyquist shaped signals were generated with optical, analog electrical and digital filters. First, the noise performance of all signals was tested using a single optical carrier. Finally, the impact of inter-channels guard bands was investigated in an ultra-densly wavelength division multiplexed scenario.

Evaluation of the Error Vector Magnitude Metric for Optical Communications: The error vector magnitude (EVM) is well known and widely employed to express the signal quality in wireline and wireless communications. Here, the applicability of the EVM for optical communications has been investigated. More precisely, it has been experimentally shown that a bit error ratio (BER) can be well estimated from a measured EVM if the optical link is operated in its linear regime [J26][J27][C14][C35][C42].

Nyquist Signaling and OFDM for Optical Access Networks: An optical access network has been designed and implemented that uses either OFDM or Nyquist signals located on several electrical carriers. An experimental demonstration showed how a broadband downstream signal is only partially processed by the colorless and sourceless subscriber units (no local laser). In addition it was shown how an uplink is realized with a centralized carrier (seed) distribution [C8][C20][C21][C24].

1 Software-defined Optics: An Overview

The capacity of today's communication networks needs to significantly increase in order to keep up with the growing demand in capacity per user on the one hand, and with the exploitation of new regional markets on the other hand. Main drivers for high per-user capacity are services such as video on demand, social networks, or cloud computing. As one of the major providers of networking and communications technology, the US based company Cisco Systems periodically gives forecasts on the development of world-wide internet protocol (IP)-traffic. The most recent forecast from May 2012 [1] is shown in Fig. 1.1.

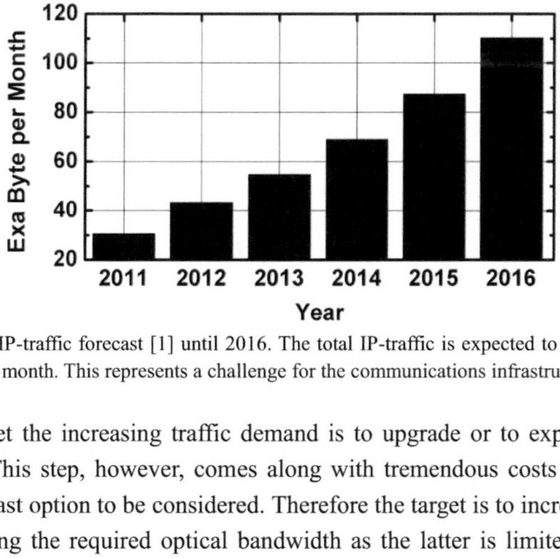

Fig. 1.1 Cisco IP-traffic forecast [1] until 2016. The total IP-traffic is expected to exceed 100 Exa (10^{18}) Bytes per month. This represents a challenge for the communications infrastructure.

One way to meet the increasing traffic demand is to upgrade or to expand the existing fiber network. This step, however, comes along with tremendous costs and is therefore most likely the last option to be considered. Therefore the target is to increase the capacity without increasing the required optical bandwidth as the latter is limited mainly by the installed optical amplifiers.

Software-defined optical transmitters (Tx) and receivers (Rx) are expected to meet this requirement as they are able to generate and receive spectrally efficient signals. In addition, their capability to adapt to changing situations in the network allows to utilize the available network resources most efficiently. As another advantage, software-defined optics shares the same hardware (HW) platform and the functionality is determined by software (SW) only. This allows to design and build a common HW for Tx and Rx transmitting various optical signals of different formats and pulse-shapes. Because of using a single HW platform only, the development- and financial efforts can be potentially decreased.

The following section gives a brief summary on non-coherent and coherent optical communications along with analog optical processing. It further shows how software-defined optics uses sophisticated digital signal processing to overcome limitations imposed by the optical transmission medium. The overall network utilization can be enhanced through the versatility of the transmitters and receivers. Finally, an overview over multi-gigabit real-time transmitters and receivers is given.

1.1 History and Vision

Originally, optical communication links were non-coherent. The most common modulation format, which is still widely employed especially in low-budget links, is the so-called on-off-keying (OOK) [2]. Here information is encoded onto the intensity of a laser, where a '0' corresponds to low (or no) intensity of light and a '1' corresponds to a higher intensity. The OOK-format allows building optical links with a rather low amount of optical components and is therefore attractive for low-cost and short-reach solutions such as access networks [3]. Unfortunately for OOK, only one bit per symbol can be transmitted which ultimately limits the spectral efficiency (SE), i.e., the amount of bits per second that are transmitted within a certain bandwidth [4]. Furthermore, since the intensity of the light is used, the maximum reach is limited by accumulated chromatic dispersion (CD) in the optical fiber [5] as well as a rather low receiver (Rx) sensitivity [6]. Limitations imposed by CD can be mitigated electronically even for optical networks with direct detection [7]. However, a perfect CD compensation can only be done for coherent optical links.

To overcome the limitations of OOK-transmission links, coherent communication links have become standard especially for long-haul transmission [8]. In these systems the electric field of an optical carrier (laser) is modulated, which provides some new degrees of freedom. This time the amplitude as well as the phase of the optical electric field is modulated [9]. A variety of complex modulation formats can be used to increase the number of bits encoded into each symbol, such increasing the SE. The most common class of modulation formats employed in coherent optical links is M-ary quadrature amplitude modulation (QAM) [10]. Today's backbone networks typically use quadrature phase shift keying (QPSK), which is also called 4QAM, and reach a data rate of 100 Gbit/s on a single optical wavelength [11]. Further optimization of coherent optical transmission links can be achieved by so-called pulse-shaping. Special pulses have been generated using pulse-carvers [12] or optical filters [13]. Some of these pulses were designed to increase the SE [14], others showed an increased non-linearity tolerance [15]. Coherent communication links have an advantage with respect to SE and Rx sensitivity [16], yet the overall costs in terms of HW and power consumption are higher than for OOK-systems. However, if the costs are related to the number of bits transmitted per second, then the relative costs of coherent optical links can be lower than for links with OOK.

With coherent communication links, digital signal processing (DSP) has become inevitable to properly receive the transmitted signals. Having the full optical field information at the Rx, fiber impairment mitigation that had formerly to be done optically (e.g. CD compensation through dispersion compensating fibers (DCF) [17]), can now be done in the digital domain [18]. Furthermore, pulse-shaping no longer requires the use of complex optics, but can be achieved with digital filters [19]. However, while optical filters can process data with tremendous bandwidths, the speed of electronics is more limited. This is mainly due to the finite sampling speed and the finite analog bandwidth of digital-to-analog converters (DAC) and of analog-to-digital converters (ADC).

Novel networks call for high SE and high flexibility at the same time to enable the transmission of high data rates [20]–[22]. So far, the wavelengths of optical transmission channels were located on a constant frequency grid, and therefore Tx and Rx could have been designed for a fixed bandwidth. The next optical network generation supports flexible grids, representing a challenge for the design of flexible optical Tx and Rx with spectrally efficient signals. Two pulse-shaping techniques have emerged as the main contenders in these flexible networks, namely orthogonal frequency division multiplexing (OFDM) [23] and Nyquist signaling [24]. Both methods require sophisticated DSP, but are expected to give the best performance and the highest flexibility.

Recently, spatial multiplexing (space division multiplexing, SDM) has emerged as a potential new way to increase the SE of coherent optical links [25]–[28]. Here, signals are transmitted using several spatial modes in the optical fiber. Again, sophisticated DSP is required as the spatially multiplexed links are so-called multiple-in-multiple-out (MIMO) [29] systems.

1.2 State of the art

Record data rates were demonstrated with coherent optical transmission and advanced DSP employing either OFDM [30], or transmitting Nyquist signals [31]. When most of the DSP for optical communications was performed offline, i.e., data are sampled, stored, and processed on conventional computers. Real-world networks call for real-time processing. Recently, several groups started to develop and demonstrate real-time Tx and Rx. Due to the lower complexity, the majority of the groups worked on the Tx only. Fig. 1.2 gives an overview of data for real-time OFDM Tx (filled circles) and Rx (open circles) and Nyquist sinc-pulse Tx (filled squares) with respect to maximum bit rate on a single polarization.

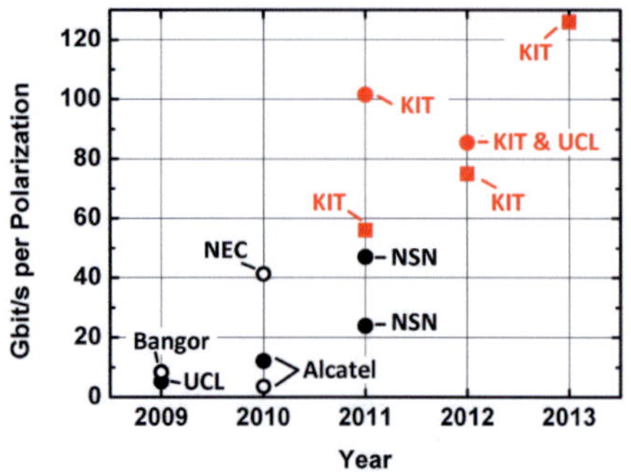

Fig. 1.2 Bitrates (per polarization) of real-time OFDM Tx (filled circles) and Rx (open circles) or Nyquist sinc-pulse Tx demonstrations (filled squares) of several groups over the years 2009–2013.

In 2009, the Optical Networks Group of the University College London (UCL) demonstrated a real-time OFDM Tx with a maximum bit rate of 8.36 Gbit/s [32]. In the same year Bangor University showed the first real-time OFDM link with Tx and Rx at a maximum bit rate of 5.25 Gbit/s [33]. One year later, in 2010, Alcatel-Lucent demonstrated a real-time OFDM of Tx at 12.1 Gbit/s [34] and a real-time OFDM Rx at 3.55 Gbit/s [35]. At the same time, NEC was able to show an OFDM Rx receiving up to 41.25 Gbit/s data in real-time [36]. In 2011, the group of Nokia Siemens Networks (NSN) implemented a real-time OFDM Tx that is able to generate 23.9 Gbit/s [37] or even 46.9 Gbit/s [38]. Also in 2011, the real-time OFDM Tx developed in the course of this thesis was demonstrated and was the first to break the 100 Gbit/s barrier with a total bit rate of 101.5 Gbit/s [39]. A little bit later, the first real-time sinc-pulse Tx was successfully demonstrated generating data up to 56 Gbit/s [40]. This Tx has been enhanced to achieve a maximum bit rate of 75 Gbit/s in 2012 [41]. In the same year, a collaboration between UCL and our group lead to an experiment with a real-time OFDM Tx capable of generating 85.4 Gbit/s [42][43]. In this year, our real-time Nyquist sinc-pulse Tx was further enhanced, and now achieves a bitrate of 126 Gbit/s [44] using the very same hardware as described in [40] and in [41].

2 Optical Transmission Systems and Related DSP

Data transmission over optical fiber has become essential for today's back-bone networks, and is increasingly employed also for access networks. The major drivers for optical communications are the low losses and tremendous bandwidth offered by optical fibers. With optical links, thousands of kilometers can be bridged while keeping the overall costs at an economic level. At the same time high performance digital circuits are now available to significantly enhance the performance of optical transmission through multi-gigabit digital signal processing (DSP). The schematic of an optical point-to-point communication link is pictured in Fig. 2.1. In this example, signals are received coherently. The link can be separated into three groups, namely the transmitter (Tx), the transmission link (or channel), and the receiver (Rx).

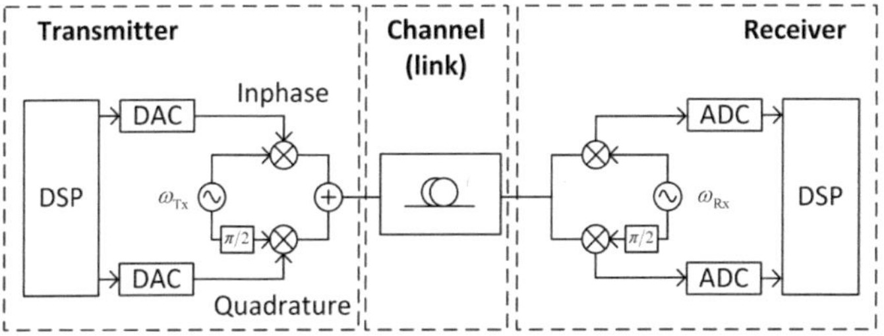

Fig. 2.1 Coherent transmission link comprising transmitter (Tx), channel (link), and receiver (Rx). In the Tx, a digital signal processing (DSP) block processes data that is fed to digital-to-analog converters (DAC). The DACs drive an I/Q-modulator to encode the electrical waveforms onto the inphase (I) and the quadrature (Q) of a carrier with frequency ω_{Tx}. The transmission link generally comprises several spans of optical fiber and optical amplifiers. In the Rx another carrier with frequency ω_{Rx} serves as a local oscillator (LO). With the LO and the I/Q-demodulator, the transmitted signal is converted to the baseband. These signals are sampled by two analog-to-digital converters (ADC) before another DSP circuit performs further processing.

Transmitter: Data is first processed by a DSP circuit and then fed to two digital-to-analog converters (DAC). The DAC outputs encode the inphase (I) and quadrature (Q) of a carrier with frequency ω_{Tx}. This is achieved by a complex optical mixer or I/Q-modulator. The signal is then launched into the transmission fiber.

Channel (link): An optical transmission channel generally comprises several spans of optical fiber. Optical amplifiers overcome the losses introduced by the fiber spans.

Receiver: At the Rx, a second carrier with frequency ω_{Rx} serves as a local oscillator (LO). Together with the I/Q-demodulator, the signal is back-converted to the baseband. The resulting electrical signals are sampled by analog-to-digital converters (ADC). Further processing is performed by another DSP circuit.

All components in Fig. 2.1 that are used for the optical transmission system are discussed in the following sections. First an overview over the optical physical layer is given. Finally, DSP techniques are discussed that enable coherent optical communications at highest speeds.

2.1 The Optical Transmitter

Encoding of an electrical signal onto an optical carrier can be achieved by several means. The most straightforward technique is to modulate the drive current of a laser source. Doing so, the optical intensity at the laser output is proportional to the electrical drive signal. This is referred to as intensity modulation (IM). However, the bandwidth of so-called directly modulated lasers (DML) is limited and the signal quality is affected by impairments such as relative intensity noise and frequency chirp. External modulators help to overcome the limitations of DMLs. Electro-absorption modulators (EAM) offer high bandwidth and are in a first approximation insensitive towards polarization changes of the incoming light. As a drawback, only intensity modulation is supported by EAMs. For full versatility at high bandwidths, so-called Mach-Zehnder interferometers are employed to modulate a continuous wave (cw) laser source. A more detailed discussion is given in the following section.

2.1.1 Mach-Zehnder Modulator

A Mach-Zehnder modulator (MZM) comprises two optical couplers and two electro-optic modulators (EOM) that apply an optical phase shift that is proportional to an applied voltage $v(t)$. The phase shifters are based on the linear electro-optic effect [45]. They are also called Pockels cells. There are different materials that can be used. One of the most common materials is $LiNbO_3$ [46]. Mach-Zehnder modulators are polarization sensitive, i.e., only light in a specific state of polarization (SOP) is modulated whereas light in the orthogonal "wrong" SOP is either absorbed in the substrate or passed to the modulator output un-modulated. The schematic of a MZM is depicted in Fig. 2.2.

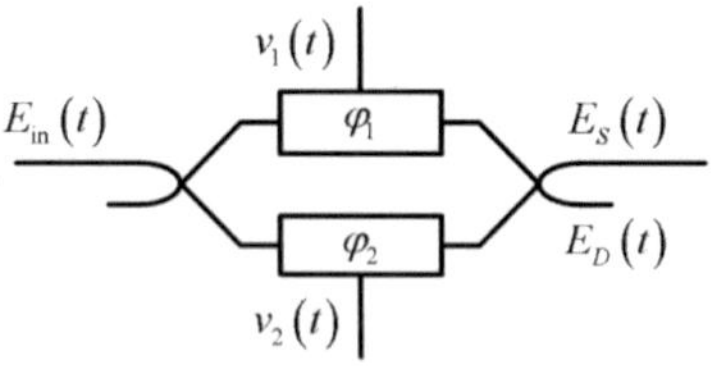

Fig. 2.2 Mach-Zehnder modulator. An optical electric field $E_{in}(t)$ is split to propagate in two arms. In each arm a phase shift φ, that is proportional to a drive voltage $v(t)$, is applied to the field. These phase shifters employ the linear electro-optic effect. Finally the fields in the two arms are combined to form the outputs $E_S(t)$ and $E_D(t)$. Usually only one of the two outputs is connected in commercial modulators.

An optical electric field can be expressed in form of an analytic signal by a "slowly varying" and generally complex amplitude $A(t)$ of an optical carrier with angular frequency ω.

$$E(t) = A(t)\exp(j\omega t) \tag{2.1.1}$$

The amplitude $A(t)$ is slowly varying in a sense that it is constant over multiple periods of the optical carrier. With an electric field $E_{in}(t)$ according to Eq. (2.1.1) the optical electric fields at the two outputs of the MZM can be described as follows:

$$E_S(t) = jE_{in}(t)e^{j\frac{\varphi_1+\varphi_2}{2}}\cos\left(\frac{\varphi_1-\varphi_2}{2}\right) \tag{2.1.2}$$

$$E_D(t) = jE_{in}(t)e^{j\frac{\varphi_1+\varphi_2}{2}}\sin\left(\frac{\varphi_1-\varphi_2}{2}\right) \tag{2.1.3}$$

Usually only one of the two outputs described by Eq. (2.1.2) and Eq. (2.1.3) is connected in commercial modulators. There are three principle modes of operation for a MZM, i.e., it can be operated as an intensity, an amplitude or as a phase modulator. If configured in so-called push-pull mode, i.e., if $\varphi_1 = -\varphi_2$, then the MZM performs pure amplitude modulation, and Eq. (2.1.2) reduces to

$$E_S = jE_{in}(t)\cos(\varphi). \tag{2.1.4}$$

As the MZM amplitude modulator has a sinusoidal transfer function, non-linear distortions of the electrical drive signals $v(t)$ are observed. For small signal amplitudes, however, $\sin(x) \approx x$ holds and the modulator is operated in its linear regime. For large signal amplitudes $v(t)$, pre-distortion can be applied to linearize the transfer function [47]. This technique comes at the price of advanced electrical drivers or, if performed digitally, the need for digital-to-analog converters (DAC) with an increased resolution.

If a pure phase modulation is desired, the modulator has to be operated in push-push mode. In this case the phase shifters are driven such that $\varphi_1 = \varphi_2$ holds and the output field is expressed by

$$E_S(t) = jE_{in}(t)e^{j\varphi} \tag{2.1.5}$$

Most MZMs are operated as amplitude modulators as in Eq. (2.1.4), and some of them are already configured to be in push-pull mode from the beginning. In that case only one external voltage $v(t)$ has to be applied which reduces the complexity of the driving electronic circuits and of the electric/optic interfaces within the modulator. By applying a direct current (DC) bias voltage to the phase shifter v_{bias} one can distinguish between amplitude and intensity modulation.

$$E_S = jE_{in}(t)\cos(\varphi(t) + \varphi_{bias}). \tag{2.1.6}$$

For amplitude modulation where the driving signal $v(t)$ is proportional to the optical field, a bias of $\varphi_{bias} = \pi/2$ is to be applied. The bias is set to the so-called null point of the modulator. In the IM case, a bias phase shift is chosen where half of the optical intensity at the input is transmitted. In this case $\varphi_{\text{bias}} = -\pi / 4$ must be applied and the modulator is operated in the quadrature point.

2.1.2 Optical I/Q-Modulator

An optical I/Q-modulator consists of two MZMs nested in a third MZM structure. These two MZMs are called children whereas the third MZM structure is referred to as parent. The schematic setup is shown in Fig. 2.3. Again an optical electric field described by Eq. (2.1.1) is split and enters the two child MZMs. These MZMs are operated as amplitude modulators and the bias phases are set to $\varphi_{\text{bias}} = \pi / 2$. In the lower arm of the parent MZM a 90° or $\pi / 2$ phase shift with respect to the optical carrier is applied. With this phase shift the so-called in-phase signal $v_I(t)$ and the quadrature phase signal $v_Q(t)$ are made orthogonal and can be separated without crosstalk given that an appropriate Rx is employed.

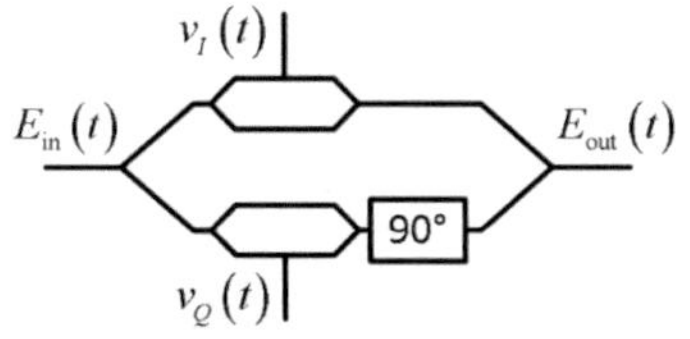

Fig. 2.3 Optical I/Q-modulator. Two MZMs are arranged in a third parent-MZM structure. The signal encoded by the two child-MZMs are referred to as in-phase (I) and quadrature phase (Q) signals, respectively. The third parent-MZM establishes orthogonality between I and Q by applying a phase shift of 90° or $\pi / 2$ with respect to the optical carrier to the quadrature arm.

With the results obtained from Eq. (2.1.6) and using analytic signals to describe the optical electric fields, the output field $E_{\text{out}}(t)$ follows from [48],

$$E_{\text{out}}(t) = \frac{E_{\text{in}}(t)}{2}\left[\cos\left(\frac{\pi}{2}\frac{v_I(t)+V_{I,\text{bias}}}{V_\pi}\right) + \mathrm{j}\cos\left(\frac{\pi}{2}\frac{v_Q(t)+V_{Q,\text{bias}}}{V_\pi}\right)\right] \tag{2.1.7}$$

In this equation the bias voltages are denoted by $v_{I,\text{bias}}$ and $v_{Q,\text{bias}}$, respectively. One of the most interesting characteristics (next to the modulation bandwidth) of a MZM is described by the V_π parameter. It basically states which voltage has to be applied to the electrodes of the modulator in order to achieve a phase shift of π. This parameter thus reveals the sensitivity of the modulator with respect to electrical signal amplitude $v(t)$. Usually a small V_π is desired as high electrical signal amplitudes especially at high frequencies in the GHz regime are not easily provided. These high amplitudes require sophisticated electrical amplifiers that will degrade the signal quality due to electronic noise.

The electric field measured at the output of the I/Q-modulator is obtained by taking the real part of Eq. (2.1.7) according to

$$\begin{aligned}\Re\{E_{\text{out}}(t)\} &= \Re\left\{\frac{|E_{\text{in}}|\mathrm{e}^{\mathrm{j}\omega t}}{2}\left[\cos(I)+\mathrm{j}\cos(Q)\right]\right\} \\ &= \frac{|E_{\text{in}}|}{2}\left[\cos(\omega t)\cos(I)-\sin(\omega t)\cos(Q)\right], \quad I,Q=\frac{\pi}{2}\frac{v_{I,Q}(t)+V_{I,Q,\text{bias}}}{V_\pi}\end{aligned} \tag{2.1.7}$$

2.2 Optical Transmission Links

Today, the majority of communication links depend on optical fiber as a transmission medium. Fibers are used due to their low losses and tremendous bandwidth. Fibers generally differ in size, geometry, and material. Depending on the application, there are glass fibers (silica) and plastic optical fibers (POF). For all types of fibers, waveguiding is achieved by a core having a high refractive index and a lower refractive index cladding. The waveguiding mechanism is based on total internal reflection [49]. Some fibers are designed such that only a single transverse mode is guided at the wavelength region of interest. These fibers are called single mode fibers (SMF). Other types of fibers support propagation of multiple modes and are therefore called multi-mode fibers (MMF). These MMF have a larger core diameter than SMF for the same wavelength region. There are also fibers designed to show special properties such as highly nonlinear fibers [50] or photonic crystal fibers [51]. Recently, so-called few-mode fibers [52] (a special case of MMF) and multi-core fibers [53] (integrating multiple SMFs into one fiber cladding) have emerged as one of the next research areas. Along with fiber, optical links also comprise amplifiers. In practice, there are three types of optical amplifiers, namely semiconductor optical amplifiers (SOA), erbium doped fiber amplifiers (EDFA), and Raman amplifiers.

In order to fully exploit the potential data capacity offered by optical fiber, one needs to investigate the fiber and amplifier characteristics. The next section covers linear fiber characteristics and briefly discusses the three mentioned optical amplifier types as they are most commonly used in today's optical networks.

2.2.1 Chromatic Dispersion

A major effect that limits the transmission of broadband optical signals in fiber is chromatic dispersion. Dispersion describes the effect that the phase velocity of an optical wave is frequency dependent. The phase velocity v_p with which a wave travels in a medium is determined by the vacuum speed of light c and the medium's refractive index n leading to

$$v_p = \frac{c}{n}. \tag{2.2.1}$$

Similar to this phase velocity of Eq. (2.2.1), the so-called group velocity v_g expresses how fast the envelope $A(t)$ of a group of waves travels. The envelope $A(t)$ represents the modulation of the wave. The group velocity v_g is then defined as the derivative of the angular frequency ω with respect to the propagation constant k:

$$v_g = \frac{\mathrm{d}\omega}{\mathrm{d}k} = \frac{c}{n_g}. \tag{2.2.2}$$

Here the group refractive index n_g is defined according to

$$n_g = n + \omega \frac{\mathrm{d}n}{\mathrm{d}\omega}, \tag{2.2.3}$$

and λ is the corresponding wavelength in vacuum. In an optical fiber, guided light propagates in only one direction. Therefore the propagation vector reduces to its z-component which is referred to as β in the following.

$$k_z = \beta \tag{2.2.4}$$

If the angular frequency ω is small compared to the angular frequency ω_c of the optical carrier, then the propagation constant β can be expanded in a Taylor series centered around ω_c according to

$$\beta(\omega_{\mathrm{opt}}) \approx \beta_c^{(0)} + (\omega_{\mathrm{opt}} - \omega_c)\beta_c^{(1)} + \frac{(\omega_{\mathrm{opt}} - \omega_c)^2}{2!}\beta_c^{(2)} + \ldots, \tag{2.2.5}$$

$$\beta_c^{(i)} = \left.\frac{\mathrm{d}^i \beta(\omega)}{\mathrm{d}\omega^i}\right|_{\omega=\omega_c}. \tag{2.2.6}$$

In Eq. (2.2.5) the first term $\beta^{(0)}$ describes a phase delay t_p whereas the second term $\beta^{(1)}$ corresponds to a group delay t_g. Both, phase and group delays, do not affect the form of the slowly varying envelope $A(t)$. Only terms of $\beta^{(2)}$ and higher order lead to a spreading or generally deformation of $A(t)$. This effect is referred to as group velocity dispersion (GVD).

If only the second derivative of the propagation constant is considered while all higher terms are neglected, then the effect of CD is quantitatively described by the so-called dispersion coefficient according to

$$C = \frac{2\pi c}{\lambda^2}\beta_c^{(2)}. \tag{2.2.7}$$

For typical standard single mode fibers (SSMF) the dispersion parameter for transmission in the C-band (1530 nm–1565 nm) is approximately 16 ps / (nm km). The deformation of the envelope $A(t)$ can be described in the baseband by an all-pass filter with a quadratic phase response. If the group delay t_g is disregarded and using Eq. (2.2.5) and Eq. (2.2.7), then the transfer function of this all-pass filter is defined by

$$G(z,\omega) = \exp(-\mathrm{j}\beta(\omega)z) = \exp\left(-\mathrm{j}C\frac{\lambda^2\omega^2}{4\pi c}z\right). \tag{2.2.8}$$

Here, λ is the wavelength of the optical carrier, z is the length of propagation, c is the speed of light in vacuum, and C is the dispersion coefficient according to Eq. (2.2.7). The inverse Fourier transform then yields the corresponding time response g(z,t)

$$g(z,t)=\sqrt{-\mathrm{j}\frac{c}{C\lambda^2 z}}\exp\left(\mathrm{j}\frac{\pi c}{C\lambda^2 z}t^2\right). \tag{2.2.9}$$

Knowing the properties of CD, the deformation of the pulse form can be undone by signal processing which will be explained in Section 2.4.3.

2.2.2 State of Polarization

Light that propagates in optical fiber generally has transverse and longitudinal field components. For weakly guiding waveguides, however, the optical electric field can be approximated by scalar quantities for the transverse x and y-direction, respectively, whereas the propagation is usually described along the z-direction. Depending on the dominant oscillation direction of the electric field E, one can distinguish between x-polarized light and y-polarized light. These two states of polarization (SOP) are orthogonal, i.e., waves with an arbitrary SOP can be described by two so-called linear polarized waves (e.g. one in x, the other in y-direction). Linear polarized light in an angle of 45° and −45° are referred to as L+ and L− polarizations, respectively. Another set of orthogonal SOPs is represented by so-called right- and left-handed circular polarized light (RC, and LC). The direction of rotation is determined face on looking at the propagating wave. Again all intermediate SOPs can be described by the superposition of these orthogonal SOPs. To visualize the polarization, Fig. 2.4 depicts the electric fields E_x and E_y for several different SOPs. Orthogonal SOPs represent two independent communication channels. This fact is exploited to double the capacity of a fiber optic channel by means of polarization division multiplexing (PDM) [54]. However, in real systems, the two orthogonal polarization channels experience crosstalk due to coupling effects. Additional impairments experienced by PDM-signals are caused by polarization mode dispersion (PMD) [55] and polarization dependent loss (PDL) [56] both of which will be explained later.

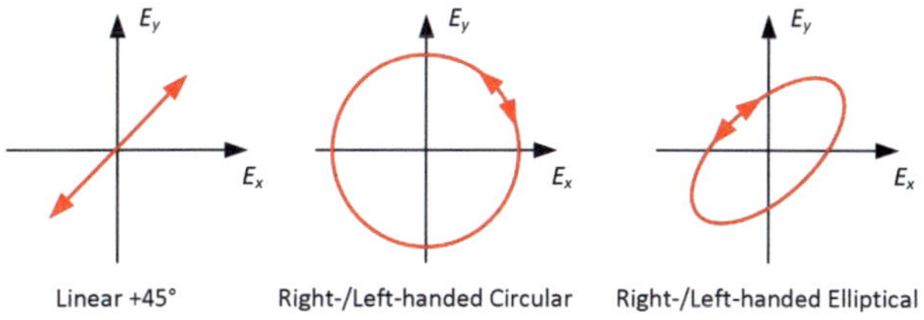

Fig. 2.4 Polarization of the electric field E. One can distinguish between linear, circular, and as something in-between elliptical polarized light.

A mechanism to describe fully polarized light is the so-called Jones calculus. The electric field E is essentially a vector comprising a component E_x with oscillation in x-direction and a component E_y with oscillation in y-direction. Replacing the complex slowly varying amplitude $A(t)$ in Eq. (2.1.1) by the modulus of the electric field E_0, the electric field vector can be described with the help of the so-called Jones vector

$$\vec{E} = \begin{pmatrix} E_x \\ E_y \end{pmatrix} = E_0 \mathrm{e}^{\mathrm{j}(\omega t - kz)} \begin{pmatrix} J_x \\ J_y \end{pmatrix}, \tag{2.2.10}$$

where

$$\vec{J} = \begin{pmatrix} J_x \\ J_y \end{pmatrix} = \mathrm{e}^{\mathrm{j}\delta} \begin{pmatrix} \mathrm{e}^{-\mathrm{j}\varphi} \cos\alpha \\ \mathrm{e}^{+\mathrm{j}\varphi} \sin\alpha \end{pmatrix}, \quad 0 \le \alpha \le \frac{\pi}{2}, \quad 0 \le \varphi < \pi . \tag{2.2.11}$$

Here α describes the magnitude of E_0 in each polarization, and φ denotes the phase offset between x and y.

- Linear polarized light: $\varphi = 0, \pi, 2\pi, \ldots$
- Circular polarized light: $\varphi = \pi / 2, 3\pi / 2, \ldots$

While the Jones calculus only describes fully polarized light, the Stokes parameters also take partially polarized light into account. The so-called Stokes vector is defined by

$$\vec{s} = \begin{pmatrix} s_0 \\ s_1 \\ s_2 \\ s_3 \end{pmatrix} = \begin{pmatrix} I \\ pI \cos 2\psi \cos 2\chi \\ pI \sin 2\psi \cos 2\chi \\ pI \sin 2\chi \end{pmatrix}, \tag{2.2.12}$$

with the total intensity $I = |E_x|^2 + |E_y|^2$ of the electric field and the degree of polarization p ($0 \le p \le 1$). Furthermore, ψ and χ are angles to determine the polarization ellipse. The normalized Stokes vector is obtained by using the intensity I (the Stokes component s_0) as reference according to

$$\vec{S} = \frac{s_1}{s_0}\vec{e}_1 + \frac{s_2}{s_0}\vec{e}_2 + \frac{s_3}{s_0}\vec{e}_3 \tag{2.2.13}$$

The normalized Stokes parameters in Eq. (2.2.13) can be interpreted as the spherical coordinates of the SOP and thus can be visualized using the so-called Poincaré sphere which is depicted in Fig. 2.5. The Stokes vector of fully polarized light ends at the surface of the sphere, whereas the Stokes vector for unpolarized light has zero length. Orthogonal SOPs can be found on opposite ends of any given Stokes vector.

In practice, optical fibers show birefringence, i.e., the refractive index depends on the polarization direction of travel. Again there are two orthogonal axes called the slow and the fast axis. These two axes are called eigenstates. If a homogeneous medium is excited with light in one of these eigenstates, it experiences no change in polarization along propagation independent of its frequency ω. When modeling coupling processes in optical fiber, then the fiber is divided into several slices rotated by arbitrary angles with respect to each other. These eigenstates of the individual slices are then a local property. In practice, there are many reasons for the variation of local eigenstates such as mechanical stress or temperature gradients.

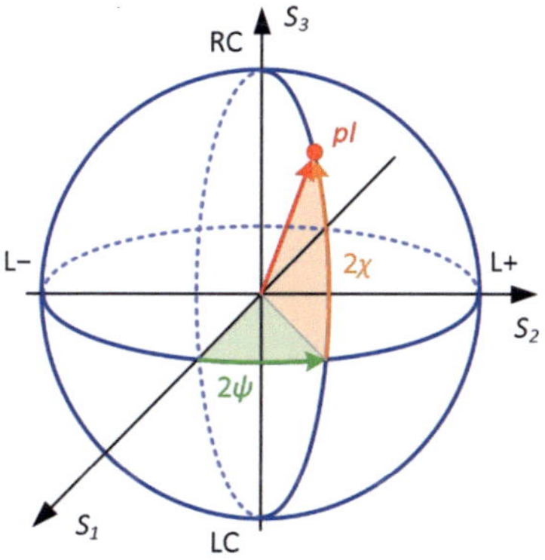

Fig. 2.5 Poincaré sphere used to visualize the SOP described by Stokes parameters.

Opposed to eigenstates, the two principle states of polarization (PSP) describe a global property of the transmission medium. These special SOP's are approximately invariant for changes in the launched signal's wavelength (or frequency ω) to first order. If excitation is not done in the principle state, then the Stokes vector shows a right-handed precession around the principle state as a function of ω.

Different polarizations travel at different speeds due to the polarization-dependent refractive index leading to so-called polarization mode dispersion (PMD). This causes a differential group delay (DGD) and results in pulse broadening similar to chromatic dispersion does. However, due to random coupling of the two launched polarizations in the fiber, the effective PMD strength scales with the square-root of propagation length. This is in contradiction to the effect of CD whose strength scales with the propagation length. Crosstalk between the two orthogonal polarizations is caused by coupling due to impurities in the fiber.

Another effect that has to be taken into account is polarization dependent loss (PDL). Many optical devices are generally polarization sensitive, i.e., the impact of the device on each of the polarizations differs. When the PDL of state-of-art optical fibers is sufficiently low and therefore negligible, other frequently used active and passive optical devices such as amplifiers or couplers may show significant PDL. Polarization-dependent loss causes the PSPs not to be orthogonal anymore, and thus significantly complicates and disturbs the demultiplexing of PDM signals that by definition rely on orthogonal channels.

As the SOP at the receiver is usually unknown and subject to change, distinct signal processing has to be performed in order to properly receive PDM signals. The incoming light is split by a polarization beam splitter (PBS) and fed to two 90° hybrids. The outputs of the PBS are essentially a superposition of two linear polarized waves.

2.2.3 Optical Amplifiers

Due to the exponential decay of transmitted signals in optical fiber, transmission distances beyond 100 km can only be achieved with optical amplifiers in-between. The two main characteristics of optical amplifiers are the gain G and the noise figure NF. For optical amplifiers (if not for phase-sensitive amplification where even signal regeneration is possible) there is a theoretical achievable minimum NF = 3dB [57]. This is opposed to electrical amplifiers that can have a theoretical minimum NF = 0 dB. Essentially, there are three non-phase-sensitive types of amplifiers that can operate in the C-band.

Semiconductor Optical Amplifier (SOA): Semiconductor optical amplifiers are similar to semiconductor lasers but without the cavity. Electrical current is used to pump a so-called gain medium. If light travels through the gain medium, amplification is obtained by stimulated emission. The SOA is the most economic optical amplifier with small size and cost while offering reasonable gain and noise figure. It further offers the potential to be integrated with other optical devices such as modulators and switches. Typical SOAs that have been used in the experiments of this work show a gain of $G \approx 14$ dB and a noise figure of NF ≈ 5 dB. There are, however, SOAs with a maximum gain of $G \approx 25$ dB [58].

Erbium Doped Fiber Amplifier (EDFA): Erbium doped fiber amplifiers are the most commonly deployed optical amplifier type today. The gain window of EDFAs covers the C-band as well as the L-band (1570 nm – 1610 nm). These amplifiers are optically pumped at either 980 nm or at 1480 nm. The 980 nm-pumped EDFAs are used as pre-amplifiers showing high gain and a low noise figure. As a drawback of pre-amplifiers, the saturation output power P_{sat} is lower than for the so-called high power amplifiers or boosters. These EDFAs are usually pumped at 1480 nm and are designed to have a high P_{sat}. In this work solely pre-amplifiers have been used that have a gain of $G \approx 40$ dB, a noise figure of NF < 4 dB, and a saturation power of $P_{sat} \approx 18$ dBm at the output. EDFAs are less energy efficient than SOAs due to the optical pumping. They also tend to be more expensive and bigger in size but usually give the best performance and therefore are often the first choice.

Raman Amplifier: Raman amplification utilizes stimulated Raman scattering which uses a pump light that transfers energy to the co-propagating optical signals by means of Raman scattering. Typically, a so-called Raman fiber serves as gain medium. As an advantage over SOAs and EDFAs where the gain is usually applied locally in a device, Raman amplification can be achieved remotely as inline distributed gain. The optical bandwidth of this mechanism is determined by the transmission window of the fiber and centered at the wavelength of the pump laser. Using multiple pumps at different wavelength enable a broadband amplification.

2.3 The Optical Receiver

The last device in an optical transmission link is the receiver (Rx). The Rx demodulates and decodes transmitted data. In optical communications one can generally distinguish between direct detection and coherent detection. The first detection scheme is simple and straightforward, but the latter scheme outperforms a direct detector in terms of sensitivity. The following sections deal with both reception schemes and discuss them in detail.

2.3.1 Direct Detection

The simplest form of an optical receiver comprises a single photodetector or photodiode (PD). With this device optical signals are converted into a photocurrent which is proportional to the intensity and to thus to the power P of the optical field. The photocurrent $i(t)$ of a PD follows from [59]:

$$i(t) = RP. \tag{2.3.1}$$

The current is proportional to the real part of the mean squared (ms, with respect to the optical carrier) optical electric field $E(t)$. The responsivity R describes the conversion efficiency of the PD. It is defined by [59]

$$R = \frac{e\eta}{\hbar\omega_s}. \tag{2.3.2}$$

Here e denotes the elementary charge and η is the quantum efficiency. In the divisor, there is the product of the reduced Planck constant $\hbar$ and the optical angular frequency ω_s. Some disadvantages of a direct detection scheme are that no polarization division multiplexing [54] can be employed, and that only real baseband signals can be received. This is in contrast to a coherent scheme where complex signals on two orthogonal polarizations are usually received.

2.3.2 Coherent Reception

For a coherent Rx not only the signal is required but a second optical electric field needs to serve as a local oscillator (LO). The signal field $E_s(t)$ can again be described by the product of a slowly varying amplitude $A_s(t)$ and the optical carrier at ω_s

$$E_s(t) = A_s(t)\exp(\mathrm{j}\omega_s t + \theta_s). \tag{2.3.3}$$

In analogy the optical electric field of the LO reads

$$E_{\mathrm{LO}}(t) = A_{\mathrm{LO}}\exp(\mathrm{j}\omega_{\mathrm{LO}} t). \tag{2.3.4}$$

A 180° optical coupler in combination with two PDs is used to construct the coherent receiver. The setup is depicted in Fig. 2.6. The signal field and the LO field are fed to either of the two coupler inputs.

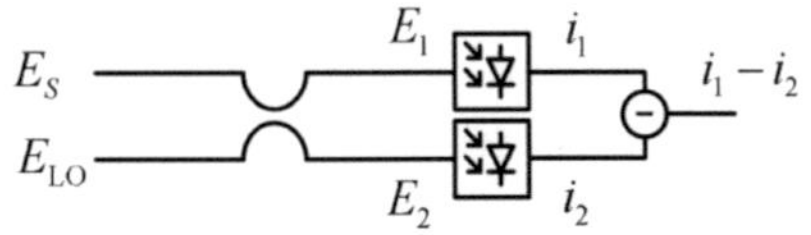

Fig. 2.6 Coherent detection scheme comprising an 180° optical coupler and a balanced PD. Two optical fields at the input (signal and LO) mix and lead to two photocurrents that are subtracted from one another to form the output photocurrent.

The two PDs see either the optical field E_1 or E_2 at their inputs. These field comprise both the signal and the LO [59].

$$E_1 = \frac{1}{\sqrt{2}}(E_s + E_{LO}) \tag{2.3.5}$$

$$E_2 = \frac{1}{\sqrt{2}}(E_s - E_{LO}) \tag{2.3.6}$$

It should be noted that both $E_s(t)$ and $E_{LO}(t)$ must occupy the same SOP in order to form the output described by Eq. (2.3.5) and Eq. (2.3.6). After quadratic detection by the two PDs and with $P \sim |E|^2$, the two output photo currents $i_1(t)$ and $i_2(t)$ follow from Eq. (2.3.1).

$$i_1(t) = RP_1 \sim \frac{R}{2}\left[|E_s|^2 + |E_{LO}|^2 + 2|E_s||E_{LO}|\cos\left(\omega_{IF}t + \theta_s(t) - \theta_{LO}(t)\right)\right] \tag{2.3.7}$$

$$i_2(t) = RP_2 \sim \frac{R}{2}\left[|E_s|^2 + |E_{LO}|^2 - 2|E_s||E_{LO}|\cos\left(\omega_{IF}t + \theta_s(t) - \theta_{LO}(t)\right)\right] \tag{2.3.8}$$

Here P denotes the power of the electric field E and ω_{IF} is the frequency difference between the signal frequency ω_s and LO frequency ω_{LO}. There are two phase terms ($\theta_s(t)$ and $\theta_{LO}(t)$) that account for the phases of signal and LO, respectively. In the two photocurrents described by Eq. (2.3.7) and Eq. (2.3.8) only the mixing products of LO and signal fields are of interest. In order to suppress remaining terms so-called balanced PDs provide a current which is the difference of $i_1(t)$ and $i_2(t)$. This output current reads [59]

$$i(t) = i_1(t) - i_2(t) \sim 2R|E_s||E_{LO}|\cos\left(\omega_{IF}t + \theta_s(t) - \theta_{LO}(t)\right) \tag{2.3.9}$$

In Eq. (2.3.9) it can be seen that the terms $|E_s|$ and $|E_{LO}|$ have vanished whereas the part of interest, i.e. the mixing product, doubles. With this kind of coherent Rx, as in the case for direct detectors, only real baseband signals can be demodulated. As clear advantage over a direct detection scheme, however, the resulting electrical output signal scales with the electric field strength of the LO. With this amplification, input signals E_s of significantly little power can still be detected leading to high Rx sensitivity irrespective of the noise of subsequent electrical amplifiers. If signals carrying complex data, e.g. modulated with quadrature amplitude modulation (QAM), need to be detected, a so-called 90° optical hybrid is to be employed as is discussed in the following.

2.3.2.1 90° Optical Hybrid and Balanced Photodetectors

A 90° optical hybrid comprises two 180° optical couplers and a quarter wave plate that introduces a 90° phase shift with respect to the optical carrier. The setup is schematically shown in Fig. 2.7. Two balanced PDs with two photodiodes each detect the electrical fields $E_1 - E_4$ at the outputs of the couplers.

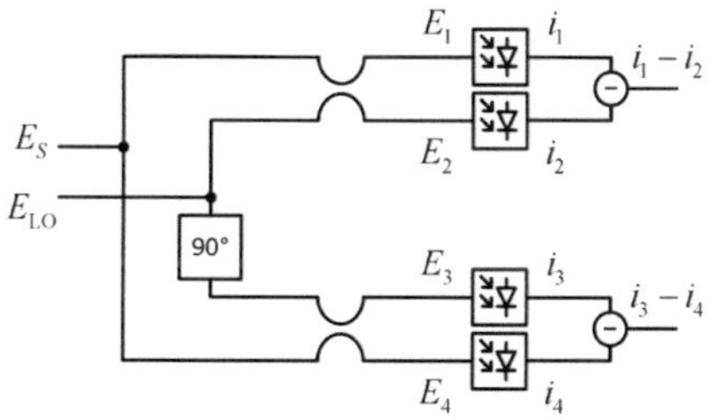

Fig. 2.7 Coherent detection scheme comprising a 90° hybrid and two balanced PDs. With this setup complex encoded optical signals are down-converted to electrical baseband signals. For PDM signals, two 90° hybrids are used where the signal is split and the LO is separated into its x and y polarization.

These electrical fields are described by the following relations [59]:

$$E_1(t) = \frac{1}{2}(E_s + E_{\mathrm{LO}}) \tag{2.3.10}$$

$$E_2(t) = \frac{1}{2}(E_s - E_{\mathrm{LO}}) \tag{2.3.11}$$

$$E_3(t) = \frac{1}{2}(E_s + \mathrm{j}E_{\mathrm{LO}}) \tag{2.3.12}$$

$$E_4(t) = \frac{1}{2}(E_s - \mathrm{j}E_{\mathrm{LO}}). \tag{2.3.13}$$

The photocurrent provided by the upper coupler is called inphase (I) component, because E_s and E_{LO} leave this coupler without any phase shift. Since the LO experiences a phase shift of 90° before entering the lower coupler, the resulting photocurrent is referred to a quadrature (Q) component of the signal. The two resulting photocurrents are described by [59]

$$i_I(t) = i_1(t) - i_2(t) \sim 2R|E_s||E_{\mathrm{LO}}|\cos(\omega_{\mathrm{IF}}t + \theta_s(t) - \theta_{\mathrm{LO}}(t)) \tag{2.3.14}$$

$$i_Q(t) = i_3(t) - i_4(t) \sim 2R|E_s||E_{\mathrm{LO}}|\sin(\omega_{\mathrm{IF}}t + \theta_s(t) - \theta_{\mathrm{LO}}(t)) \tag{2.3.15}$$

With $i_I(t)$ and $i_Q(t)$ the complex signal can be recovered according to [59]

$$i_C(t) = i_I(t) + \mathrm{j}i_Q(t) \sim 2R|E_s||E_{\mathrm{LO}}|\exp\left(\mathrm{j}\left[\omega_{\mathrm{IF}}t + \theta_s(t) - \theta_{\mathrm{LO}}(t)\right]\right). \qquad (2.3.16)$$

If information has been encoded on the phase of the Tx laser (e.g. for quadrature phase shift keyed (QPSK) signals) evaluation of Eq. (2.3.16) provides this information. The phase information of the transmitted signal $\theta_s(t)$ can be decomposed in the phase of the Tx laser $\theta_{\mathrm{Tx}}(t)$ and the phase encoded data $\theta_{\mathrm{Data}}(t)$ wheras $\theta_{\mathrm{LO}}(t)$ represents the phase of the LO. For PDM signals, two 90° optical hybrids are used. Typically the signal is fed to a 3dB optical coupler, and the LO is split into its x and y polarization using a so-called polarization beam splitter (PBS). Depending on the intermediate frequency (IF) ω_{IF} of the two optical fields, one can distinguish between three different cases of coherent reception. These scenarios are discussed in the following.

2.3.2.2 Homodyne Reception

Homodyne reception provides an electrical output signal which is in the baseband since signal carrier and LO have the same optical frequency, see Fig. 2.8(a). The coherent Rx comprises a 90° hybrid and balanced PDs according to Fig. 2.7. As the resulting electrical signal is in the baseband, see Fig. 2.8(b), the intermediate angular frequency ω_{IF} is zero. If LO and signal carrier are phase locked and have the identical absolute phases, i.e. $\theta_{\mathrm{Tx}}(t) = \theta_{\mathrm{LO}}(t)$, the complex output signal reduces to [59]

$$i_C(t) \sim 2R|E_{\mathrm{LO}}||E_{\mathrm{Data}}|\exp\left(\mathrm{j}\theta_{\mathrm{Data}}(t)\right) \qquad (2.3.17)$$

In this scenario neither frequency offset compensation nor carrier phase recovery has to be taken care of in the Rx digital processing, and data can be extracted right away.

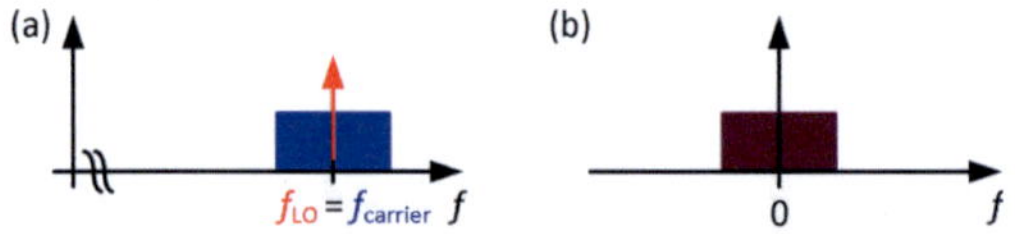

Fig. 2.8 Homodyne reception. (**a**) The signal carrier and the LO have the same optical frequency $f_{\mathrm{LO}} = f_{\mathrm{carrier}}$. (**b**) The coherently received signal is in the baseband.

2.3.2.3 Intradyne Reception

For intradyne reception ω_{IF} is non-zero, since the frequencies of signal carrier and LO are only approximately same, see Fig. 2.9(a). However, this frequency offset is usually chosen as small as possible (given that independent laser sources are used) in order to keep the required bandwidth of the Rx at minimum requiring the intermediate frequency to be considerably smaller than the signal bandwidth. For reception, a 90° hybrid is commonly employed (as in Fig. 2.7) to recover a complex valued signal representing a frequency shifted version of the transmitted signal, see Fig. 2.9(b). Intradyne receivers need to provide slightly more electrical bandwidth than homodyne Rx as they not only have to cover the signal bandwidth but also the frequency mismatch between Tx laser and Rx laser. In practice, most coherent systems use intradyne reception, since the two independent laser sources can only be locked in frequency with considerable effort [60]. For state-of-the-art tunable external cavity lasers (ECL) the frequency difference can be adjusted to be below 200 MHz. That way only little more bandwidth has to be offered by a intradyne Rx compared to a homodyne Rx.

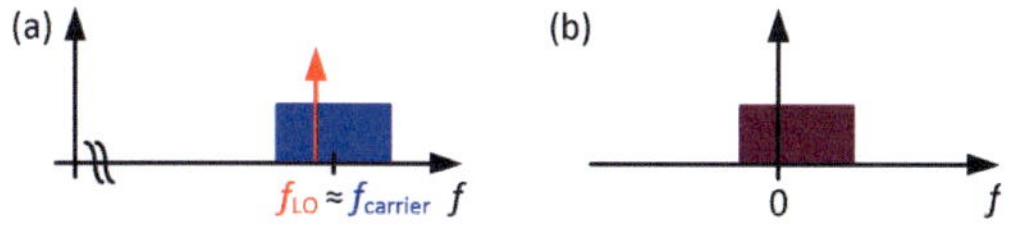

Fig. 2.9 Intradyne reception. (**a**) Signal carrier and LO have optical frequencies that are close. (**b**) The coherently received signal is shifted by the intermediate frequency between signal carrier and LO.

2.3.2.4 Heterodyne Reception

Heterodyne receivers often use an LO having a frequency offset ω_{IF} from the signal carrier at least 1.5 times larger than the signal bandwidth, see Fig. 2.10(a). As an advantage over homodyne and intradyne reception, a heterodyne Rx produces real signals only. Therefore a 90° hybrid is not needed and the setup shown in Fig. 2.6 is sufficient. The large frequency offset is chosen so that signal-signal mixing (blue triangle in Fig. 2.10(b)) does not impair the signal of interest (purple rectangle in Fig. 2.10(b)). The main disadvantage is the significantly increased need for large electrical bandwidth.

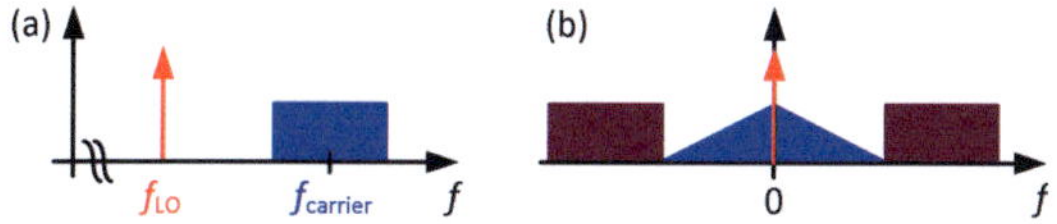

Fig. 2.10 Heterodyne reception. (**a**) The LO frequency is at least 1.5 times the signal bandwidth off the signal carrier. (**b**) Using a single square-law PD, the signal (spectrum: purple rectangle) can be fully recovered. Carrier-carrier (red line) and signal-signal mixing (blue triangle) do not impair the signal quality because of the large frequency offset between LO and signal carrier.

2.4 Digital Signal Processing for Optical Communications

Knowing the physical properties of optical modulators, fiber-optic channels, and optical demodulators, sophisticated DSP circuits for optical Tx and Rx can be designed to significantly enhance the performance or even enable the use of coherent optical communication links. An overview of the Tx and Rx-DSP on the physical layer level is given in Fig. 2.11. In this schematic, transmission and coherent reception of polarization division multiplexed signals are considered.

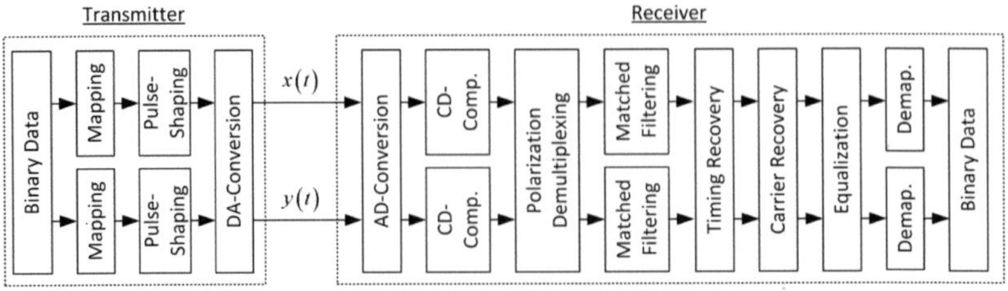

Fig. 2.11 Schematic overview of Tx and Rx-DSP on the physical layer. **Tx:** Binary data is split in *x*- and *y*-polarization information before it is mapped to complex symbols according to the modulation format. Pulse-shaping then allows for transmission of spectrally efficient signals before the DA-conversion translates digital data to analog waveforms. Two complex valued signals $x(t)$ and $y(t)$ are transmitted on two orthogonal polarizations. **Rx:** After AD-conversion accumulated chromatic dispersion (CD) is undone. This can be done for the two polarizations individually. A polarization demultiplexing block separates the *x*- and *y*-signals before a matched filter removes unwanted spectral components. With the timing recovery, the Rx is synchronized to the Tx clock. The carrier recovery then estimates the carrier frequency and carrier phase of the signals. A linear equalizer is employed to compensate for the non-flat transfer function of the analog electronics. Finally, after demapping the binary data is obtained.

Transmitter: Binary data is split into information to be transmitted in *x*- and in *y*-polarization, respectively. A mapper translates the binary data to complex symbols that are chosen according to the modulation format (e.g. *M*-ary QAM). Since both polarizations are independent, different formats can be used. Digital pulse-shaping is performed to generate signals that are e.g. spectrally efficient or show other advantages such as high non-linearity tolerance. Finally, the digitally generated signals are fed to the digital-to-analog (DA) converters where analog waveforms $x(t)$ and $y(t)$ are created to drive optical modulators

Receiver: The received analog waveforms are analog-to-digital (AD) converted to process the data digitally. As a first step accumulated chromatic dispersion (CD) is compensated. This can be done for the $x(t)$ and $y(t)$ signals individually which may pose an advantage when using multiple digital circuits at the Rx. Next polarization demultiplexing is performed to separate the two signals such that minimum inter-polarization interference (IPI) is observed. Matched filters optimize the quality of the received signals by removing unwanted spectral components such as neighboring channels in frequency or wavelength division multiplexed transmission. In this case, the inter-carrier intereference (ICI) is minimized. The timing recovery then enables synchronization of the Rx to the Tx clock and thus minimizes inter-symbol interference (ISI). As a next step, the carrier needs to be recovered. Therefore the carrier frequency and the carrier phase is estimated. A linear equalizer finally compensates for the non-flat frequency response of the analog electronics before the complex symbols are demapped and the binary data are obtained.

In the following a brief description of quantized and time discrete signals is presented before the major DSP blocks in Fig. 2.11 are discussed in detail. Therefore the upcoming section includes existing algorithms that are widely employed in optical communications as well as novel concepts that have been developed in the course of this thesis.

2.4.1 Quantized and Time Discrete Signals

Processing in the digital domain is often preferred due to the versatility and tremendous processing power of state-of-the-art digital circuits. To do so, an analog waveform $u_{\text{analog}}(t)$ is sampled at equidistant times. In order to obtain a full digital representation $u_{\text{sampled}}(t)$ of the analog waveform $u_{\text{analog}}(t)$, the sampling interval or sampling period Δt must be chosen such that the Nyquist-Shannon sampling theorem is satisfied [61][62]. For a bandwidth-limited signal with highest frequency component f_{max}, the sampling theorem is fulfilled if a minimum of either two real-valued samples or one complex-valued sample is taken within one period $1/f_{\text{max}}$. Thus the minimum sampling rates $f_s = 1/\Delta t$ for real and complex sampling follow:

$$f_{\text{s, real}} \geq 2 f_{\text{max}}, \qquad f_{\text{s, complex}} \geq f_{\text{max}} \tag{2.4.1}$$

If the condition of Eq. (2.4.1) is violated, so-called alias frequencies appear in the sampled signal. The process of sampling can be expressed as a multiplication of the signal $u_{\text{analog}}(t)$ with a Dirac pulse-train with the Dirac pulses each separated by Δt [63].

$$u_{\text{sampled}}(t) = \Delta t \sum_{n=-\infty}^{+\infty} u_{\text{analog}}(t)\delta(t - n\Delta t) \tag{2.4.2}$$

The Fourier transform of a Dirac pulse-train with spacing Δt is a Dirac pulse-train in frequency with spacing $f_s = 1/\Delta t$. The multiplication in Eq.(2.4.2) is replaced by a convolution in the frequency domain and the resulting signal spectrum of the sampled signal becomes periodically repeated to infinity:

$$U_{\text{sampled}}(f) = \sum_{n=-\infty}^{+\infty} U_{\text{analog}}(f - nf_s) \tag{2.4.3}$$

An illustration of the sampling theorem is given in Fig. 2.12. Two scenarios are shown where the sampling theorem is either satisfied (Fig. 2.12(a) and (b)) or violated (Fig. 2.12(c) and (d)). If the theorem is satisfied, the analog waveform $u_{\text{analog}}(t)$ can be reconstructed without error using a so-called reconstruction low-pass. Assuming an ideal low-pass filter, the reconstruction is equal to an interpolation of the sampled signal with a sinc-function. This is also known as the Whittaker–Shannon interpolation formula [64]. If the theorem is violated, portions of the periodically repeated image spectra fall into the main spectrum. The images overlap. An error-free reconstruction of the signal is no longer possible.

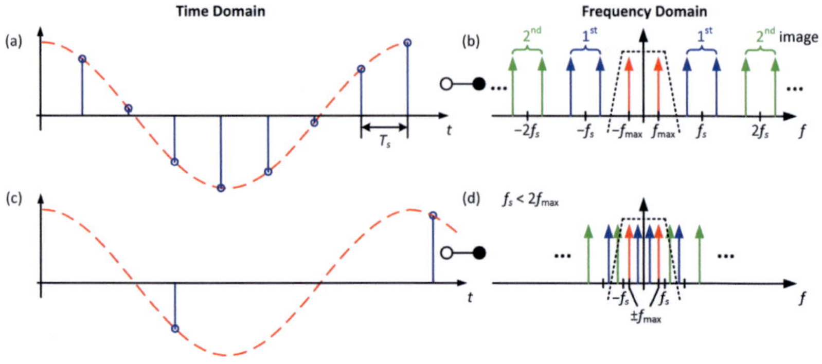

Fig. 2.12 (**a**) Time continuous waveform $u_{analog}(t)$ (here a cosine-function) sampled at equidistant times. The sampling period is $\Delta t = 1/f_s$ and satisfies the Nyquist-Shannon sampling theorem. (**b**) Main spectrum (red) along with periodically repeated image spectra (blue and green) of the sampled signal $u_{sampled}(t)$. Image spectra do not overlap with the main spectrum and the signal can thus be reconstructed by an appropriate low-pass filter (dotted frequency response). (**c**) Waveform $u_{analog}(t)$ (again the same cosine-function as in (b)) sampled at a rate f'_s lower than required by the sampling theorem. It can be seen, that there are less than two real samples within one period of the cosine-function. (**d**) The signal spectrum shows clear overlap between image spectra (blue and green) due to the violation of the sampling theorem. The original signal $u(t)$ can no longer be reconstructed.

In addition to the time discrete character of $u_{sampled}(t)$, digital circuits process quantized signals, i.e., signals with a discrete and finite number of different amplitude levels. Thereby, the quantization depth is determined by the number of resolution bits of the digitizer (e.g. an ADC). The difference between the actual analog signal amplitude $u_{analog}(t)$ and the projected digital amplitude level $u_{quantized}(t)$ is referred to as quantization error and results in quantization noise [63]. Fig. 2.13(a) shows a quasi-analog waveform (resolved with 32 bit double precision) and its quantized counterpart. The resolution is 5 bit corresponding to only 32 distinct amplitude levels.

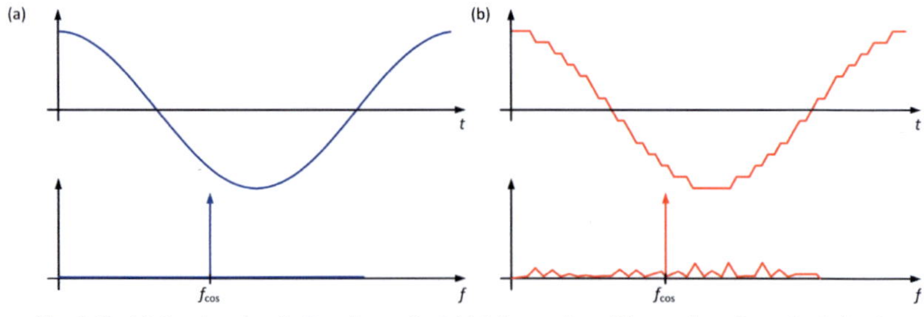

Fig. 2.13 (**a**) Quasi-analog (left) and quantized (right) waveform. The number of amplitude levels for the quantized waveform is 32 corresponding to a resolution of 5 bit. (**b**) Associate signal spectra reveal the effect of quantization noise (right) which reduces the SNR of the quantized signal.

The spectra of these signals in Fig. 2.13(b) illustrate the effect of quantization noise. The SNR of the quantized signal is significantly reduced as compared to the quasi-analog signal.

2.4.2 Complex Modulation

In general, information can be encoded onto a carrier by modulating a physical property of that carrier. This could be the amplitude, the phase, the frequency, or the polarization of a carrier with center frequency $\omega = 2\pi/T$. If an I/Q-modulator is used, such as discussed in Section 2.1.2, the complex information can be encoded onto the amplitude and phase phase of the carrier. At the Rx, the inphase (I, real part of the modulation) and the quadrature (Q, imaginary part of the modulation) of the carrier is detected, compare Section 2.3.2.1. There is no crosstalk between I and Q as the following orthogonality relation holds,

$$\frac{1}{T}\int_{-T/2}^{+T/2} \sin(\omega t)\cos(\omega t)\,\mathrm{d}t = 0\,. \tag{2.4.4}$$

The generally complex modulation coefficients can be described by

$$c = |c|\mathrm{e}^{\mathrm{j}\theta}\,. \tag{2.4.5}$$

As an example, the constellation diagrams of two commonly used modulation formats, namely QPSK and 16QAM, are depicted in Fig. 2.14. Using the QPSK format, Fig. 2.14(a), two bit of information can be transmitted per symbol whereas for 16QAM, Fig. 2.14(b), four bit are encoded into a symbol.

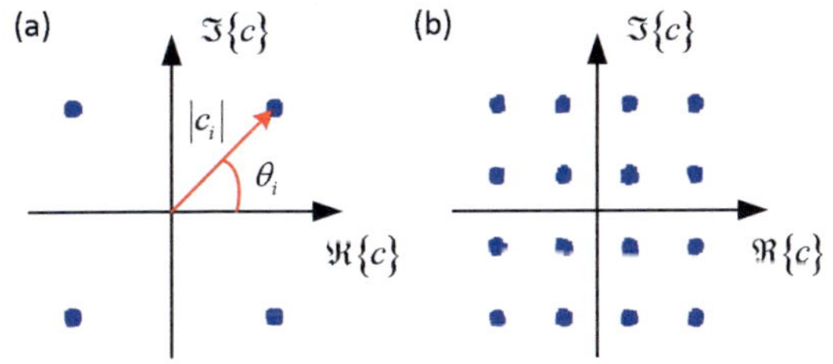

Fig. 2.14 Complex modulation formats with complex coefficients c_i where the real part is modulated onto the inphase (I) of the carrier and the imaginary part is modulated onto the quadrature (Q) of the carrier. (**a**) Constellation diagram of a qudrature phase shift keyed (QPSK) signal. (**b**) Quadrature amplitude modulated (QAM) signal constellation with a modulation alphabet of 16 symbols (16QAM).

2.4.3 Chromatic Dispersion Compensation

Since the effect of chromatic dispersion (CD) is linear, see Section 2.2.1, it can be compensated for by linear digital filters either in the Tx (precompensation) [65] or in the Rx

[66]. Digital dispersion compensation is an alternative to so-called dispersion compensating fibers (DCF), that are used to undo accumulated dispersion after each transmission span [67]. Disadvantages of optical dispersion compensation are the usually higher loss and non-linearity of the DCF as compared to standard single mode transmission fiber. A disadvantage of digital dispersion compensation is, that the computational effort scales with the amount of dispersion that is to be undone. Digital filters have been demonstrated which are able to compensate for 15,000 ps/nm of dispersion in real-time [66].
As discussed in Section 2.2.1, the effect of CD on a signal can be described by an all-pass filter with a quadratic phase response in the frequency domain, Eq. (2.2.8). In order to undo accumulated dispersion, the sign of the dispersion coefficient needs to be changed leading to an impulse response according to [68]

$$g(z,t)=\sqrt{\frac{\mathrm{j}c}{C\lambda^2 z}}\,\mathrm{e}^{-\mathrm{j}\phi(t)},\quad \phi(t)=\frac{\pi c}{C\lambda^2 z}t^2 \tag{2.4.6}$$

Unfortunately, the frequency response of such a filter is infinitely extended in time and can therefore only be approximated in a finite window. The real and imaginary parts of the filter's rectangularly-windowed (dashed rectangle) impulse response Eq. (2.2.9) is depicted in Fig. 2.15(a).

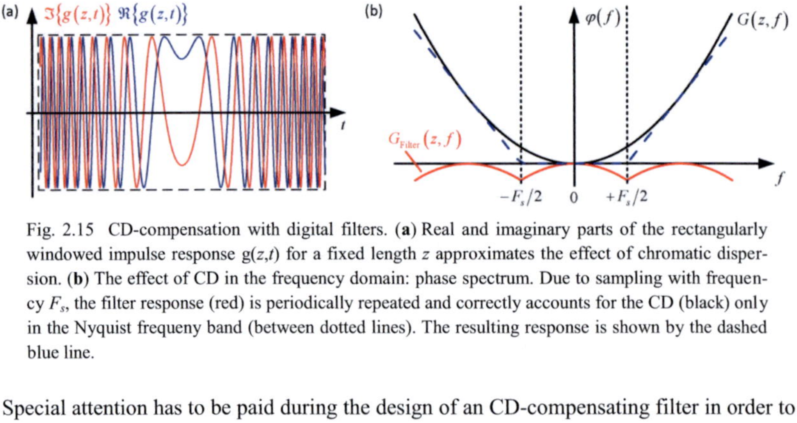

Fig. 2.15 CD-compensation with digital filters. (**a**) Real and imaginary parts of the rectangularly windowed impulse response g(*z*,*t*) for a fixed length *z* approximates the effect of chromatic dispersion. (**b**) The effect of CD in the frequency domain: phase spectrum. Due to sampling with frequency F_s, the filter response (red) is periodically repeated and correctly accounts for the CD (black) only in the Nyquist frequeny band (between dotted lines). The resulting response is shown by the dashed blue line.

Special attention has to be paid during the design of an CD-compensating filter in order to avoid aliasing. For a better understanding of the interplay between time discrete processing, see Section 2.4.1, and aliasing, the phase spectrum of a dispersive channel *G* along with the phase response of the filter is shown in Fig. 2.15(b). It can be seen that the frequency response of the dispersive channel follows a parabolic phase (black), Eq. (2.2.8) and that the frequency response of the filter is periodically repeated spaced by the sampling frequency F_s (red). The resulting phase of the dispersion compensated channel (combined transfer function of the dispersion causing channel and of the filter) is indicated by the dashed blue line. Only within the Nyquist frequency range (between the dotted grid lines), the accumulated dispersion is correctly accounted for and the phase response is flat. It has to be taken care of, that there are no signal components outside the Nyquist frequency band. This is usually achieved by the process of so-called oversampling in combination with analog anti-alias filters. A more detailed discussion is given in Section 4.2.2.

The maximum supported frequency $f_{\max}$ within a bandwidth-limited signal of a dispersion compensating filter can be taken from [68]

$$f_{\max} = \frac{c}{C\lambda^2 z} t \,. \tag{2.4.7}$$

This relation is obtained from the derivation of the phase in the exponential term of Eq. (2.2.9) with respect to time. With Δt being the minimum sampling interval according to Nyquist [61] where $f_{\max} = 1/(2\Delta t)$ and with Eq. (2.4.7) one finds

$$-\frac{C\lambda^2 z}{2c\Delta t} \le t \le \frac{C\lambda^2 z}{2c\Delta t} \,. \tag{2.4.8}$$

By obeying these conditions, a CD-compensating FIR filter of order R can be designed the coefficients of which are defined according to [68]

$$h_r = \sqrt{\frac{\mathrm{j}c\Delta t^2}{C\lambda^2 z}} \exp\left(-\mathrm{j}\frac{\pi c\Delta t^2}{C\lambda^2 z} r^2\right) \text{with } -\frac{R}{2} \le r \le +\frac{R}{2} \text{ and } R = 2\left[\frac{C\lambda^2 z}{2c\Delta t^2}\right] + 1 \,. \tag{2.4.9}$$

Since the filter order R is an integer number, the fraction of Eq. (2.4.8) in the right of Eq. (2.4.9) is rounded to the nearest integer. More detailed information about FIR filters and digital signal processing (DSP) is given in Section 2.4.7. Naturally the presented filter can also be realized in the frequency domain. In this case, the signal is transformed to the frequency domain and the convolution with the filter's impulse response is replaced by a simpler multiplication with the corresponding frequency response. Finally, the filtered signal spectrum is transformed back to the time domain.

2.4.4 Polarization Demultiplexing

Blind Polarization Demultiplexing with Low Computational Complexity

R. Schmogrow, P.C. Schindler, C. Koos, W. Freude, and J. Leuthold

IEEE Photon. Technol. Lett. 25(13), pp. 1230–1233 (2013).

Polarization division multiplexing (PDM) [54] is widely employed in coherent optical links, because it doubles the spectral efficiency and thus the transmission capacity. Together with quadrature phase shift keying (QPSK), PDM enables 100G transmission per wavelength channel [11].

To utilize data transmission in two orthogonal states of polarization (SOP), two optical transmitters (Tx) launch their signals in x and y-polarization described by complex electric field amplitudes $E_{x,t}$ and $E_{y,t}$ in x and y-direction, respectively. While propagating along optical fiber, the SOP permanently changes due to birefringence. In order to properly separate the two signals $E_{x,t}$ and $E_{y,t}$ at the receiver (Rx), the accumulated change in SOP needs to be undone. This can be done either optically [69] or by using digital signal processing (DSP) [68]. Here, we focus on DSP enabled polarization demultiplexing.

The probably most famous and widely employed polarization demultiplexing algorithm is the so-called constant modulus algorithm (CMA) [70] originally designed for phase shift keyed (PSK) signals only, but later adapted for higher order modulation formats such as M-ary quadrature amplitude modulation (QAM). The strength of the CMA is the low computational efforts that are required. As a downside, however, the algorithm needs to be implemented format specific and requires a proper timing recovery prior to execution. As an alternative, overcoming the CMA drawbacks and offering high versatility, polarization demultiplexing in Stokes space has been proposed [71]. This algorithm requires quite extensive computations, representing a challenge for real-time Rx implementations. While these two algorithms operate blindly, there is also data-aided polarization demultiplexing [72]. The data-aided algorithms make use of training symbols to separate x and y-components of PDM signals. A major disadvantage of data-aided schemes is, next to the induced overhead, that the SOP must not change between two adjacent training symbols or preambles.

In this paper, we propose and experimentally demonstrate a novel algorithm to blindly demultiplex arbitrary PDM signals with little processing efforts. Our algorithm combines the low computational complexity of the CMA with the versatility of the Stokes algorithm, i.e., it is applicable to arbitrary signals without requiring any timing recovery. We experimentally validate our algorithm with PDM-QPSK and PDM-16QAM signals and several pulse-shapes.

2.4.4.1 Novel Blind Polarization Demultiplexing Scheme

An algorithm used for demultiplexing of PDM signals needs to undo changes of the initial SOP at the Tx that are accumulated while the signals propagate in the optical fiber.

We describe these effects by an SOP transfer matrix $\mathbf{U}$, which is assumed to be unitary [71], $\mathbf{U}^\dagger\mathbf{U}=\mathbf{I}$ (unity matrix $\mathbf{I}$), i.e., the total optical power in both polarizations remains constant. The matrix $\mathbf{U}$ with complex elements u_1 and u_2 translates the optical electric field at the Tx ($E_{x,t}$ and $E_{y,t}$) to the electric field seen by the Rx ($E_{x,r}$ and $E_{y,r}$) according to

$$\begin{pmatrix} E_{x,r} \\ E_{y,r} \end{pmatrix} = \mathbf{U}\begin{pmatrix} E_{x,t} \\ E_{y,t} \end{pmatrix}, \quad \mathbf{U} = \begin{pmatrix} u_1 & u_2 \\ -u_2^* & u_1^* \end{pmatrix}, \quad u_{1,2} \in \mathbb{C}. \tag{2.4.10}$$

For our present investigation we neglect polarization mode dispersion (PMD) and polarization dependent loss (PDL), which would be important for long-haul transmission only. In the described scenario, all SOP changes can be compensated for by a concatenation of a phase retardation matrix $\mathbf{\Phi}$ and a rotation matrix $\mathbf{R}$ according to

$$\begin{pmatrix} E'_{x,t} \\ E'_{y,t} \end{pmatrix} = \mathbf{R\Phi} \begin{pmatrix} E_{x,r} \\ E_{y,r} \end{pmatrix} = \begin{pmatrix} \cos\alpha & -\sin\alpha \\ \sin\alpha & \cos\alpha \end{pmatrix} \begin{pmatrix} 1 & 0 \\ 0 & \exp(\mathrm{j}\varphi) \end{pmatrix} \begin{pmatrix} E_{x,r} \\ E_{y,r} \end{pmatrix}. \tag{2.4.11}$$

If the phase-retardation φ ($0 \le \varphi < 2\pi$) and the rotation angle α ($0 \le \alpha < \pi/2$) in Eq. (2.4.11) are chosen correctly such that $\mathbf{R\Phi U} = \mathbf{I}$, then the resulting optical electric fields components $E'_{x,t}$ and $E'_{y,t}$ do not show any inter-polarization crosstalk. To find the correct parameters α and φ, we need a simple criterion that requires minimum computational effort.

To this end, the signal powers $\langle |E'_{x,t}|^2 \rangle$ or $\langle |E'_{y,t}|^2 \rangle$ (averaged over many data samples) could be measured. We further assume that the average powers transmitted in both polarizations are equal. Maximizing (or minimizing) the appropriate powers would help if only one polarization $E_{x,t}$ or $E_{y,t}$ was used for signal transmission. However, as soon as PDM is employed and the signals in both input polarizations carry uncorrelated data $\langle E_{x,t} E^*_{y,t} \rangle = 0$, neither the sum nor the difference of the received average signal powers in orthogonal polarizations are suitable as a criterion, because we find that their values are constant irrespective of α and φ,

$$\begin{aligned} &\langle |E'_{x,t}|^2 \rangle + \langle |E'_{y,t}|^2 \rangle = 2\langle |E_t|^2 \rangle, \langle |E_{x,t}|^2 \rangle = \langle |E_{y,t}|^2 \rangle = \langle |E_t|^2 \rangle, \\ &\langle |E'_{x,t}|^2 \rangle - \langle |E'_{y,t}|^2 \rangle = 0. \end{aligned} \tag{2.4.12}$$

However, when averaged field moduli instead of powers are evaluated, the situation is less trivial, and the sum $\langle |E'_{x,t}| \rangle + \langle |E'_{y,t}| \rangle$ does reveal information on the correct parameter pair α and φ. Hence, we define a so-called cost function

$$\begin{aligned} &J_1(\alpha,\varphi) = \langle |E'_{x,t}| \rangle + \langle |E'_{y,t}| \rangle = \frac{1}{I} \sum_{i=1}^{I} \left[|E'_{i,x,t}| + |E'_{i,y,t}| \right], \\ &\text{with } |F| = \sqrt{\Re\{E\}^2 + \Im\{E\}^2}. \end{aligned} \tag{2.4.13}$$

To visualize the cost function $J_1(\alpha,\varphi)$, we use a noise-free PDM signal of sinc-shaped 16QAM pulses, and average over a block length of $I = 256$ samples for a randomly chosen fiber transfer matrix $\mathbf{U}$, see Fig. 2.16(a). The cost function is periodic in α and φ. Inside the fundamental period, two maxima are found which refer to complementary situations where the role of $E'_{x,t}$ and $E'_{y,t}$ is exchanged. For both maxima, the correct parameters α and φ, used to undo the SOP changes, can be read. We further plot $J_1(\alpha,\varphi)$ in Fig. 2.16(b), if Gaussian noise is added for a bit error ratio (BER) of 10^{-2}, close to the upper limit for soft-decision forward error correction. Independent of noise, the general shape of $J_1(\alpha,\varphi)$ remains same. It is also independent of sampling time, modulation format (QPSK to 1024QAM), and roll-off factor. For single-polarization signals, $J_1(\alpha,\varphi)$ exhibits two minima instead of the previous maxima.

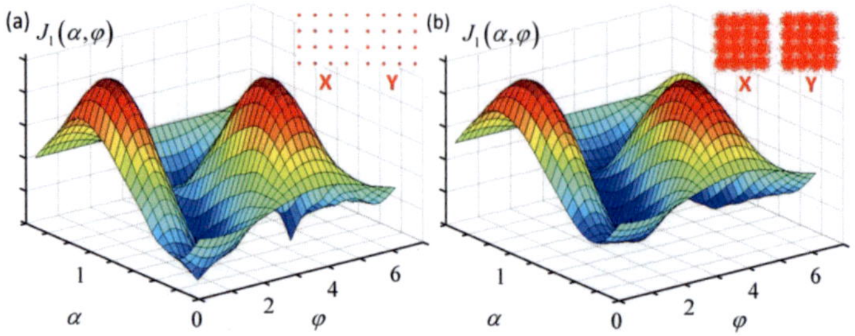

Fig. 2.16 Cost function $J_1(\alpha,\varphi)$ for PDM-16QAM signals. The maxima of $J_1(\alpha,\varphi)$ provide the correct parameters α and φ. For both maxima, the role of received x and y-polarizations are exchanged. Constellation diagrams after demultiplexing are shown as insets. (**a**) Noise-free signal. (**b**) Signal with additive Gaussian noise (BER = 10^{-2}) and identical position of maxima.

The computation of J_1 requires 4 real multiplications, 2 real square root operations, and 3 real additions for each of the I samples, see Eq. (2.4.13). Especially the square root and multiplication operations are computationally expensive. In the following, we show how this complexity can be reduced.

Simplification of the Cost Function: The largest processing effort in computing J_1 is calculating $|E|$ in Eq. (2.4.13). To study the modulus function, we depict it as a function of $\Re\{E\}$ and $\Im\{E\}$, see Fig. 2.17(a). It can be seen that it forms a parabolic "bowl". Alternatively, the sum $|\Re\{E\}|+|\Im\{E\}| \geq |E|$ is depicted in Fig. 2(b).

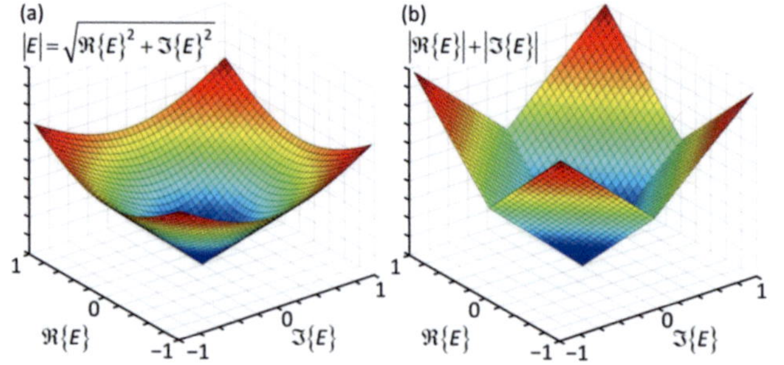

Fig. 2.17 Comparison between $|E|$ and $|\Re\{E\}|+|\Im\{E\}|$ as a function of $\Re\{E\}$ and $\Im\{E\}$. (**a**) Modulus $|E|$ forms a parabolic "bowl". (**b**) $|\Re\{E\}|+|\Im\{E\}| \geq |E|$ exhibits the same tendency as $|E|$.

It shows the same tendency as the function in Fig. 2.17(a) and suffices to find extrema in Eq. (2.4.13). Hence, a simplified cost function is

$$J_2(\alpha,\varphi) = \left\langle \left|\Re\left\{E'_{x,t}\right\}\right| + \left|\Im\left\{E'_{x,t}\right\}\right| + \left|\Re\left\{E'_{y,t}\right\}\right| + \left|\Im\left\{E'_{y,t}\right\}\right| \right\rangle$$
$$= \frac{1}{I}\sum_{i=1}^{I}\left[\left|\Re\left\{E'_{i,x,t}\right\}\right| + \left|\Im\left\{E'_{i,x,t}\right\}\right| + \left|\Re\left\{E'_{i,y,t}\right\}\right| + \left|\Im\left\{E'_{i,y,t}\right\}\right|\right]. \tag{2.4.14}$$

This simplified cost function Eq. (2.4.14) is visualized in Fig. 2.18(a) for the same noise-free signal as in Fig. 2.16(a). Again, two global maxima can be identified, leading to identical parameters α and φ. If the rotation angle α is close to 0 or $\pi/2$, local maxima appear. The feedback control is designed to avoid these local maxima, as will be explained later on. As stated for the cost function $J_1(\alpha,\varphi)$, the shape of $J_2(\alpha,\varphi)$ does not change if the same amount of Gaussian noise is added, Fig. 2.18(b). As before, sampling time, modulation format, and roll-off factor are of no consequence. For single-polarization signals, $J_2(\alpha,\varphi)$ exhibits again minima instead of the previous maxima. The computation of $J_2(\alpha,\varphi)$ involves neither multiplications nor square root operations, and needs only 3 real additions and 4 moduli of real numbers (using two's complement for negative numbers) for each of the I samples.

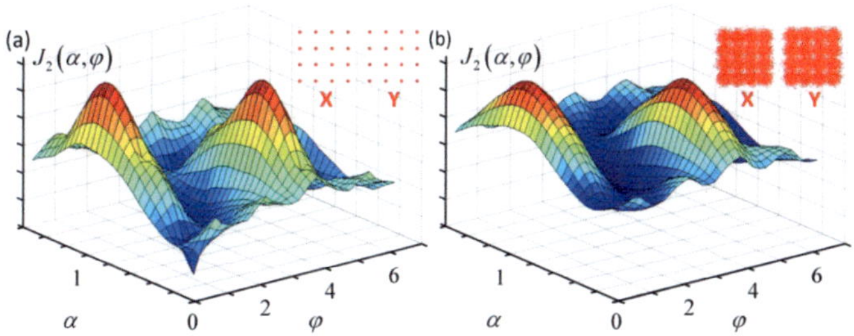

Fig. 2.18 Cost function $J_2(\alpha,\varphi)$, the global maxima of which provide the correct parameters α and φ. For both global maxima, the role of received x and y-polarizations are exchanged. Constellation diagrams after demultiplexing are shown as insets. **(a)** Result for a noise-free PDM-16QAM signal as in Fig. 2.16(a) **(b)** Adding noise to the signal leads to shallower maxima than for noise-free signals. However, this does not affect the functioning.

A schematic of the polarization demultiplexing DSP according to Eq. (2.4.11) is shown in Fig. 2.19(a). The rotation matrix $\mathbf{R}$ is realized with a standard butterfly filter. The cost function $J = J_{1,2}$ is computed according to the schematic Fig. 2.19(b). For finding the maxima, we employ a gradient ascent algorithm in a Cartesian (α,φ)-coordinate system with unit vectors $\mathbf{e}_\alpha, \mathbf{e}_\varphi$. We define the gradient of $J = J_{1,2}$ by

$$\nabla J = \frac{\partial J}{\partial \alpha}\mathbf{e}_\alpha + \frac{\partial J}{\partial \varphi}\mathbf{e}_\varphi \tag{2.4.15}$$

and approximate the derivatives by central differences. For a coarse maximum search we choose large step sizes, where the algorithm is not caught by local extrema. When a maximum is coarsely localized, the step size is refined until the algorithm converges. This happens within a few thousand symbols.

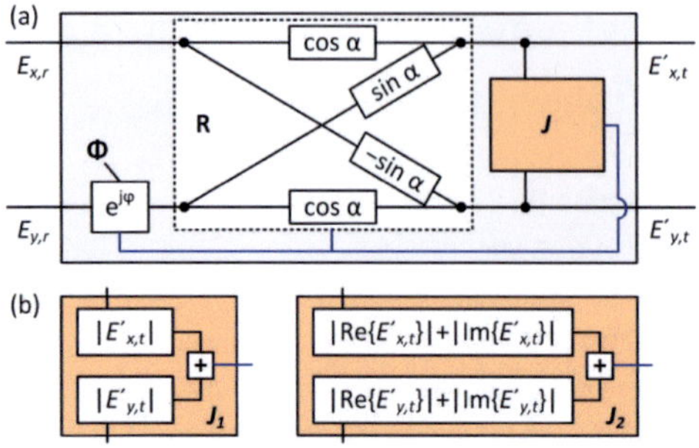

Fig. 2.19 (**a**) DSP schematic for demultiplexing of PDM signals. A phase shift of φ is applied to one of the sampled electric fields before a butterfly filter rotates both fields by an angle α. (**b**) Either the cost function J_1 or J_2 is computed to determine the optimum parameters φ and α.

The choice of the block length I is uncritical as long as all symbols are represented with approximately equal probability. Our algorithm converged for 16QAM signals with minimum block lengths of $I = 64$.

Experiments: The experimental setup is depicted in Fig. 2.20. We use a software-defined Tx [73] comprising two field programmable gate arrays (FPGA), two Micram digital-to-analog converters (DAC), and an optical I/Q-modulator to encode complex data on an external cavity laser. All waveforms are generated offline and stored in the FPGAs. The signals are split, decorrelated by 14 m of polarization maintaining fiber, and combined in orthogonal SOPs. An erbium doped fiber amplifier (EDFA) adjusts the optical power levels. A polarization scrambler (Pol.) changes the SOP randomly on an ms-scale, resulting in a virtually constant SOP within each measurement period of 25 µs. Finally, the signals are coherently received with an Agilent optical modulation analyzer for offline processing

We generate QPSK and 16QAM signals with a symbol rate of 14 GBd. To illustrate the versatility of the proposed polarization demultiplexing algorithm, we choose different pulse-shapes, including the transmission of rectangular non-return-to-zero (NRZ) pulses as well as so-called "raised-cosine" (RC) pulses with different spectral roll-off factors β [74]. No timing recovery is performed prior to polarization demultiplexing. For the PDM signals we further launch the signals in x and y-polarization either in synchronism or delayed by half a symbol duration. In practice, the latter case is often preferred as it usually leads to lower optical power fluctuations, thus increasing the non-linearity tolerance.

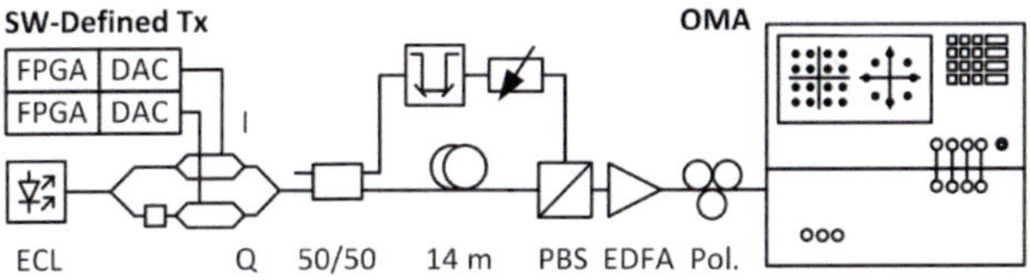

Fig. 2.20 Experimental setup. A pair of field programmable gate arrays (FPGA) and digital-to-analog converters (DAC) act as an arbitrary waveform generator. Offline generated signals are encoded on an external cavity laser with an I/Q-modulator. The signals are split, decorrelated by 14 m of fiber and combined in orthogonal SOPs using a polarization beam splitter (PBS). An erbium doped fiber amplifier (EDFA) boosts the signals. A polarization controller (Pol.) randomly changes the SOP before an optical modulation analyzer (OMA) receives the signals coherently.

First, signals on a single polarization are transmitted. The results for RC-QPSK and RC-16QAM signals can be seen in Fig. 2.21(a) and (b), which show the color-coded contour plots for the cost functions J_1 and J_2, respectively. To the right, the constellation diagrams received in x and y-polarization are displayed (upper rows, black: prior to polarization demultiplexing; bottom rows, blue: after polarization demultiplexing). The single dot in two of the constellation diagrams refers to zero field strength with noise. Due to the ambiguity of the cost functions, the total signal power can be received either in x-polarization, Fig. 2.21(a), or in the y-polarization, Fig. 2.21(b).

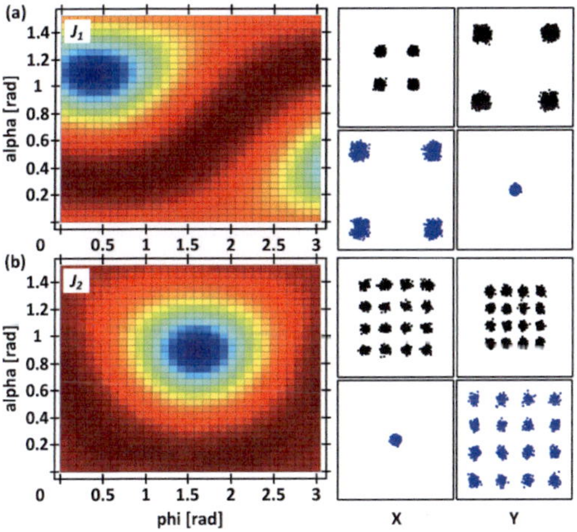

Fig. 2.21 Measured polarization demultiplexing for signals with raised-cosine spectrum on a single polarization. The minima (dark blue circular regions) of cost functions (**a**) J_1 and (**b**) J_2 yield the parameters to demultiplex (**a**) QPSK and (**b**) 16QAM signals (upper rows, black). After demultiplexing, all signal power is either in x or in y-polarization (lower rows, blue). The single dot in the constellation diagrams refers to zero field strength with noise.

Second, we transmit PDM-QPSK and PDM-16QAM signals with different pulse-shapes. The results are illustrated in Fig. 2.22. Received signals prior to polarization demultiplexing show severe crosstalk, see constellation diagrams (black) in Fig. 2.22(a)–(c) (upper rows). In addition to the polarization crosstalk, the timing recovery fails for the interleaved PDM signals in Fig. 2.22(b), (c). The evaluation of the cost functions J_1 or J_2 is shown by the plots left to the constellation diagrams (color-coded contour plots). Searching for a maximum of $J_{1,2}$ (dark red circular regions) yields the correct parameters α and φ for optimum polarization demultiplexing. This is possible even for very small signal-to-noise ratios as was demonstrated in Fig. 2.16(b) and Fig. 2.18(b). Constellation diagrams after polarization demultiplexing (red) are depicted in Fig. 2.22(a)–(c) (lower rows). We experimentally tested the proposed techniques for NRZ, Fig. 2.22(a), sinc-like pulse-shapes with RC spectra and a roll-off $\beta = 1$, Fig. 2.22(b), and sinc-shaped pulses with $\beta = 0$, Fig. 2.22(c). The QAM formats and pulse-shapes chosen for Fig. 2.22 represent a selection of most common signal types. We thus conclude that neither the modulation format nor the pulse-shape influence the performance of our polarization demultiplexing scheme.

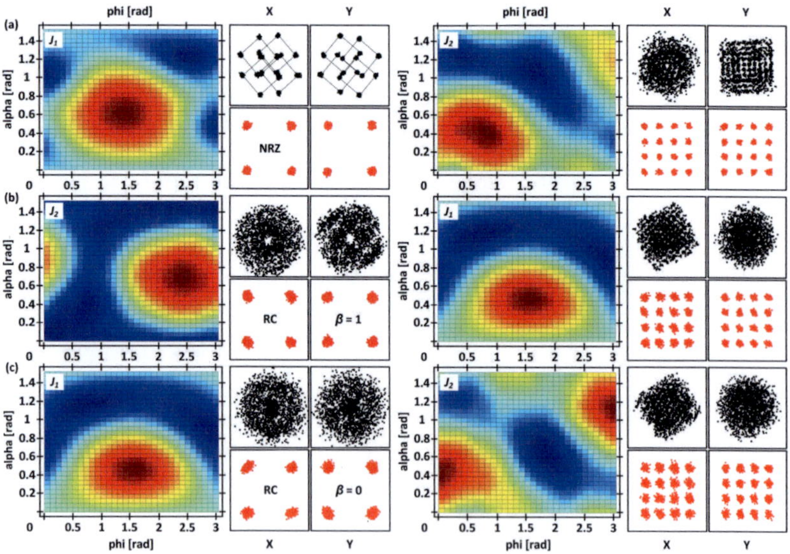

Fig. 2.22 Measured polarization demultiplexing of PDM-QPSK (left) and PDM-16QAM signals (right) with different pulse-shapes. Cost functions J_1 and J_2 (color-coded contour plots) are depicted along with measured constellation diagrams. Constellation diagrams prior to polarization demultiplexing (black, upper rows) are shown above the constellation diagrams after polarization demultiplexing (red, lower rows). **(a)** Non-return-to-zero (NRZ) pulse-shape. **(b)** Sinc-like pulse-shape with raised-cosine (RC) spectrum and roll-off factor $\beta = 1$. **(c)** Sinc-shaped pulses with spectral roll-off factor $\beta = 0$.

2.4.4.2 Polarization Demultiplexing in Stokes Space

An algorithm called demultiplexing in Stokes space has been demonstrated in [71]. The algorithm is used to separate arbitrary PDM signals. Hence, its functionality is independent of the modulation format, the signal timing, and the pulse-shape. The algorithm is based on the principle of analyzing received data in terms of unnormalized Stokes parameters to find the desired polarization transformation matrix. The main advantage of this algorithm is that polarization alignment and demodulation can be performed separately.

Starting from the complex optical electric fields E_x and E_y and using the Jones vector of Eq. (2.2.11), the resulting field E is defined by

$$E_t = \frac{1}{\sqrt{2}}\begin{pmatrix} E_x \\ E_y \end{pmatrix} = \frac{1}{\sqrt{2}}\begin{pmatrix} a_x \exp\left(\mathrm{j}\omega t + \mathrm{j}\varphi_x\right) \\ a_y \exp\left(\mathrm{j}\omega t + \mathrm{j}\varphi_y\right) \end{pmatrix}. \tag{2.4.16}$$

To transform the field described by Eq. (2.4.16) to a set of Stokes parameters as in Eq. (2.2.12) the following relation is used

$$S = \frac{1}{2}\begin{pmatrix} e_x e_x^* + e_y e_y^* \\ e_x e_x^* - e_y e_y^* \\ e_x^* e_y + e_x e_y^* \\ -\mathrm{j}e_x^* e_y + \mathrm{j}e_x e_y^* \end{pmatrix} = \frac{1}{2}\begin{pmatrix} a_x^2 + a_y^2 \\ a_x^2 - a_y^2 \\ 2a_x a_y \cos\left(\varphi_y - \varphi_x\right) \\ 2a_x a_y \sin\left(\varphi_y - \varphi_x\right) \end{pmatrix}. \tag{2.4.17}$$

As seen in Eq. (2.4.17), the advantage of this representation is, that the Stokes parameters are independent of the carrier frequency ω. Therefore residual carrier fluctuations do not influence the measurement. The outcome for a PDM-QPSK signal is shown in Fig. 2.23. It can be seen that the measured PDM-QPSK data points describe a lens in the unnormalized Stokes parameter space. This lens indicates a plane through the origin of the plot, whose normal identifies the orthogonal SOPs of the signals. It should be noted that, despite accumulated ASE noise, the signals are fully polarized at all times. Just because the computed Stokes parameters remain unnormalized, the distance of the measured points to the origin is not constant as would have been expected for normalized parameters.

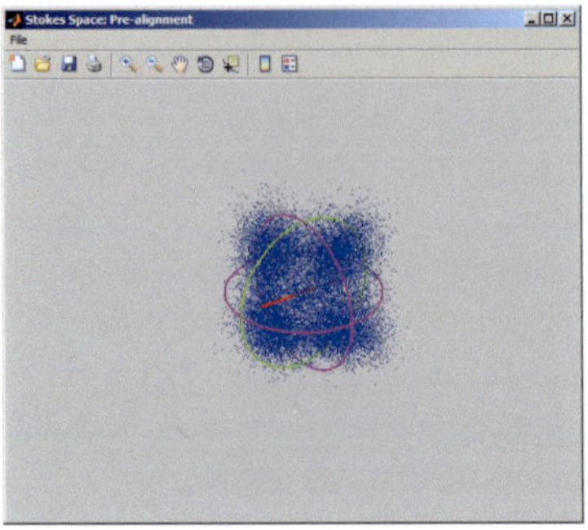

Fig. 2.23 Output of the polarization demultiplexing in Stokes space algorithm provided by the Agilent N4391A Software for a measured PDM-QPSK signal. The measured Stokes parameters help to identify a plane whose normals provide the SOP seen at the Rx.

2.4.4.3 Constant Modulus Algorithm

The constant modulus algorithm (CMA) [70] has been designed to optimize finite-duration impulse response (FIR) filters that e.g. separate signals from two orthogonal polarizations. The CMA strictly requires modulation formats that exhibit only phase variations such as QPSK or 8-PSK. However, the algorithm has been adapted to higher order modulation formats like *M*-ary QAM [75] as well. Taken e.g. a PDM-QPSK signal that is received with a PBS followed by two 90° optical hybrids and balanced photo detection (see Fig. 2.7). If the polarization axes of the Tx and Rx are perfectly aligned, then the signal on each polarization is purely phase modulated and thus has a constant modulus. If there is a misalignment between Tx and Rx, then the separated signals have multiple amplitude levels due to the superposition of tributaries belonging to different polarizations. The CMA designs a transformation matrix or filter such that the output signals are demultiplexed with minimum crosstalk. This is achieved by minimizing a so-called cost function J, i.e., minimizing the deviation from a constant modulus (in this example normalized to 1):

$$\begin{aligned} J_1 &= E\left[\left(1-|y_1|^2\right)^2\right], \\ J_2 &= E\left[\left(1-|y_2|^2\right)^2\right]. \end{aligned} \tag{2.4.18}$$

Here y is the signal at the filter output and E denotes the expectation value. According to Eq. (2.4.18), the optimization process computes the least mean squares, a method which is further discussed in Section 2.4.7. There are two main disadvantages of the CMA. First, the specific modulation scheme has to be known to the Rx and the demodulation and polarization demultiplexing processes are not independent. Second, the CMA shows a singularity, i.e., the filter might not converge to the optimal solution [59].

2.4.4.4 Data-aided Demultiplexing

When the previous two algorithms were based on blind equalization, a popular option is so-called data-aided equalization. For the latter technique essentially a known sequence is repeatedly sent which is often called preamble. With this known sequence of symbols or bits, the Rx can easily compute the polarization transfer matrix to align the Tx and Rx SOP [72]. The repetition rate of these preambles is determined by the maximum change of SOP over time. In between two preambles, the SOP should remain constant. As a major drawback, the preamble introduces overhead and thus reduces the achievable data rate. However, data-aided algorithms are often used due to their simplicity, low complexity, and guaranteed convergence.

2.4.5 Digital Filtering Techniques at the Receiver

In optical communication links often several carrier wavelengths are used to transmit independent signals in order to fully facilitate the bandwidth offered by optical fiber and optical amplifiers. Doing so, the discrepancy between broadband optics and rather narrowband electronics can be bridged. At the Rx, only the signal of interest needs to be processed while signals on neighboring wavelength channels must be filtered in order to avoid inter-channel interference (ICI). Since optical filters usually have shallow filter slopes, and as inter-channel guard-bands need to be kept as small as possible for maximum spectral efficiency (SE), digital filters with steep slopes are used to remove spurious parts of the neighboring channels. In the following two digital filtering techniques at the Rx are discussed, namely matched filtering and so-called Nyquist filtering.

2.4.5.1 Matched Filtering

The matched filter was originally known as North filter since it was proposed by Dwight O. North in 1943 [76]. In order to achieve matched filtering, the received signal is correlated with a known part of the signal or a template (e.g. an elementary pulse of a signal) in order to detect this template in the signal and thus extract the transmitted modulation coefficients c. A discrete correlation between the signal u and the matched filter's impulse response h follows from

$$y(n)=\sum_{r=-\infty}^{+\infty} h(r)u(n-r) \qquad (2.4.19)$$

The operation described by Eq. (2.4.19) can be realized by convolving the received signal with the time-reversed and complex conjugate version of the template. It can be shown, that a matched filter maximizes the signal-to-noise ratio (SNR) in the presence of additive stochastic noise [76]. It further removes signal components of neighboring channels. Unfortunately, there are some restrictions on the elementary pulse-shape or template of the matched filtered signals if ISI-free signal reception is desired. In this case not all pulse-forms are supported. So for instance so-called square-root-raised-cosine (RRC) filters [77] have a matched filter at the Rx whereas raised-cosine (RC) filters do not have a ISI-free matched filter [77]. For these and other pulse-shapes, Nyquist filtering can enhance the signal performance as will be explained in the following.

2.4.5.2 Nyquist Filtering

The following section is an excerpt from [J12].

Pulse-Shaping with Digital, Electrical, and Optical Filters – A Comparison

R. Schmogrow, S. Ben-Ezra, P.C. Schindler, B. Nebendahl, C. Koos, W. Freude, and J. Leuthold

J. Lightw. Technol. 31(15), pp. 2570–2577 (2013). [J12]

To overcome limitations introduced by components with finite electrical bandwidth, we investigate advanced Rx processing techniques such as equalizers with finite impulse response (FIR) filters to minimize ISI, or "brick-wall" digital filtering to suppress signals outside the Nyquist frequency bandwidth f_{Nyq} that equals the symbol rate (named "Nyquist filtering" in the following). In addition to Nyquist filtering we employ an advanced clock recovery scheme.

First the coherently received signals are polarization de-multiplexed [71] if polarization division multiplexing (PDM) [78] was applied. In this section we investigate signals on a single polarization only and thus omit the polarization de-multiplexing block. The remaining DSP blocks are shown in Fig. 2.24(a). The corresponding signal spectra are displayed in Fig. 2.24(b). The black spectrum on top (Tx) corresponds to electrically shaped QPSK signals as received and sampled by the Rx. It shows a significant roll-off within the pass-band because of the electrical anti-alias filters in the Tx, see inset Fig. 2.24. Due to this roll-off, the Nyquist ISI criterion is violated. The Nyquist filtering block removes the signal spectrum outside the Nyquist frequency band (Fig. 2.24(b), red spectrum NYQ), yet the Tx caused roll-off remains. The Nyquist filtering includes also the clock recovery [79]. Next, a static pre-equalizer coarsely flattens the combined Tx and Rx transfer functions hence mitigating ISI (Fig. 2.24(b), blue spectrum NYQ + $\mathrm{EQ_S}$). Finally, a blind post-equalizer ($\mathrm{EQ_D}$, using the least-mean square algorithm [80]) is adapted by evaluating the measured error vector magnitude (EVM) [81][82]. It removes any residual roll-off and thus residual ISI. Inside this control loop the carrier phase is recovered.

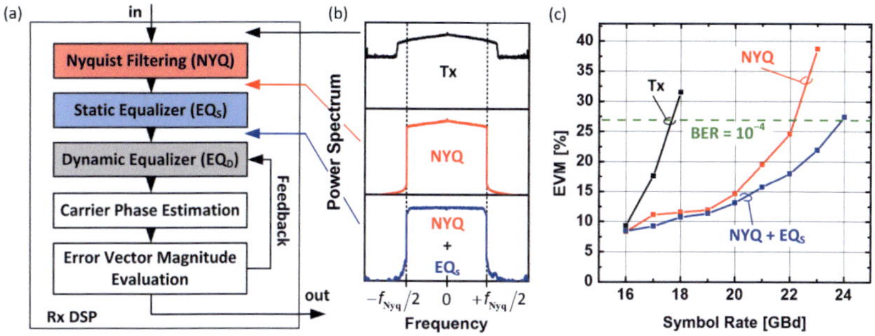

Fig. 2.24 Digital signal processing (DSP) blocks in the receiver together with the influence of the various equalization elements on the achievable symbol rate. (**a**) A Nyquist filtering block (NYQ) removes the signal spectrum outside the Nyquist frequency band and provides clock recovery. The static pre-equalizer (EQs) coarsely flattens the combined Tx and Rx transfer functions. A blind post-equalizer is dynamically adjusted by evaluating the measured error vector magnitude (EVM). Inside this control loop the carrier phase is recovered. (**b**) Ensemble averaged spectra for 20 GBd signals. The unfiltered spectrum (top, Tx) shows a significant roll-off. After Nyquist filtering portions outside the Nyquist frequency band are removed (middle, NYQ). The static equalizer (EQs) additionally flattens the Nyquist pass-band. (**c**) EVM versus symbol rate plotted for an analog Tx with cut-off frequency $f_{el} \approx 12.3$ GHz. (Black, Tx): Dynamic equalizer only. (Red, NYQ): Dynamic equalizer and Nyquist filtering. (Blue, NYQ + EQs): All DSP blocks in Subfigure (**a**) are active. The better a flat brick-wall spectrum is approximated, the closer the symbol rate approaches the Nyquist rate of 24.6 GBd.

To judge the influence of the DSP blocks preceding the dynamic equalizer EQ_D, we measured the signal quality (EVM) of the electrically shaped NRZ-QPSK as a function of the symbol rate. The LO of the coherent Rx has been tuned to approximately match the wavelength of the Tx laser (intradyne reception). For a single polarization QPSK signal we adjusted the symbol rate in 1 GBd steps from 16 GBd to 24 GBd. The Tx low-pass filters, applied to both the in-phase and quadrature of the signals, limit the analog bandwidth to a cut-off frequency $f_{el} \approx 12.3$ GHz. These Tx filters also mimic a possible Rx bandwidth limitation assuming a linear transmission system. A maximum symbol rate of 2×12.3 GBd = 24.6 GBd results [61]. The outcome is displayed in Fig. 2.24(c). Due to convergence issues, using solely the adaptive post-equalizer EQ_D, a maximum symbol rate of only 17.5 GBd can be achieved for a minimum bit error ratio of BER = 10^{-4} (black, Tx). Activating the Nyquist filtering block (red, NYQ) enhances the possible symbol rate to 22.5 GBd. If in addition the static equalizer is turned on, a maximum symbol rate of 24 GBd is found. This comes close to the theoretical limit [61]. Said Nyquist filtering and clock recovery has already been demonstrated for *M*-ary QAM as high as 512QAM [83].

2.4.6 Carrier and Timing Recovery

Timing, carrier frequency and phase recovery for OFDM and Nyquist signals using a mean modulus algorithm

R. Schmogrow, B. Nebendahl, A. Josten, P.C. Schindler, C. Koos, W. Freude, and J. Leuthold

Opt. Express 22(8), pp. 9344–9359 (2014). [J2]

Clock and carrier recovery are crucial tasks when coherently receiving data transmitted over optical links. Although a variety of synchronization algorithms are known from wireless and wireline communications, it is the use of extremely high data rates (frequently beyond 100 Gbit/s) in optical links which represents a big challenge. With the advent of computationally expensive pulse-shaping techniques like Nyquist signaling [24] [79] and orthogonal frequency division multiplexing (OFDM) [23] for optical networks, standard blind synchronization methods fail. However, Nyquist and OFDM signaling have become important as they allow for coding at highest spectral efficiency [84]. Nyquist signals comprise sinc-like pulses which overlap in time but do not show any inter-symbol interference (ISI). Such sinc-like pulses are infinitely extended in time but have a finite "raised-cosine" spectrum [74], the shape and width of which are described by the roll-off factor β. On the other hand, OFDM signals have finitely extended rect-shaped symbols with infinitely extended sinc-shaped subcarrier (SC) spectra that overlap in frequency. The SC spectra are separated by the reciprocal symbol duration, are orthogonal and hence free of inter-channel interference (ICI).

Timing recovery (also: clock recovery) and carrier recovery methods can be grouped into feedforward and feedback techniques. In both cases, the actual value of frequency, phase or sampling time has to be estimated. This information is represented by a control variable, which either leads to a feedforward correction, or gives the actual value in a feedback loop.

In the past, for single carrier formats such as Nyquist signals, a number of methods have been proposed to provide data for synchronizing the receiver (Rx) clock to the Tx clock. One way to find the proper timing information is to use square-law detectors [85] [86], i.e., the modulus of the received time domain signal is squared. However, this algorithm is not applicable for Nyquist pulses with a small roll-off factor $\beta \approx 0$ [85] [86]. A solution suitable for signals with small β is to use absolute-value rectifiers [87]. The distorted signal can then be fed to a fast Fourier transform (FFT) in order to obtain the clock frequency and phase information from the spectrum with peaks at the clock frequency [88] [89].

In a next step, carrier recovery is performed, usually using the well-known Viterbi-Viterbi algorithm [90] that requires a proper timing synchronization in advance. The Viterbi-Viterbi method provides both the carrier frequency and the carrier phase offset.

For multicarrier signals such as with OFDM, the Schmidl-Cox algorithm [91] is widely used. It computes the correlation of training patterns to provide timing and carrier frequency information at the same time.

In this paper we propose and experimentally demonstrate simple and highly efficient blind algorithms for timing information recovery as well as for carrier frequency and phase estimation. In particular, it is shown that our mean modulus algorithm (MMA) is well suited for both Nyquist and OFDM signals. It can be used to extract clock, carrier frequency and phase while featuring a low computational complexity [79] [43].

The paper is organized as follows: We first present an intuitive view on the operation principle followed by the theoretical description of the basic recovery techniques in a generalized form (applicable to both Nyquist and OFDM signals). Next, we compare the MMA to the well-known CMA. Finally, we discuss the experimental setup and apply the algorithms to OFDM signals with 128 SCs and Nyquist pulses with various roll-off factors β and data rates of up to 84 Gbit/s.

2.4.6.1 Operation Principle

In this section we give an intuitive explanation on the operation principles of the clock recovery, which is investigated in more detail further below. Exemplarily, we show the operation principle by studying the color-coded eye-diagrams and mean values (black dashed lines) of two signals with different pulse shapes as a function of time and symbol period T_s, see Fig. 2.25. The amplitude envelope of the first signal is plotted in Fig. 2.25(a), upper row. It was obtained from a binary phase shift keyed (BPSK) signal with a sinc-like pulse shape (also known as a "raised-cosine" pulse due to its raised-cosine spectrum). The amplitude envelope of the second signal is plotted in Fig. 2.25(b), upper row. This time we evaluate a BPSK signal with a Nyquist sinc pulse shape. The mean value of the amplitude envelopes is a straight line, and no timing information can be derived.

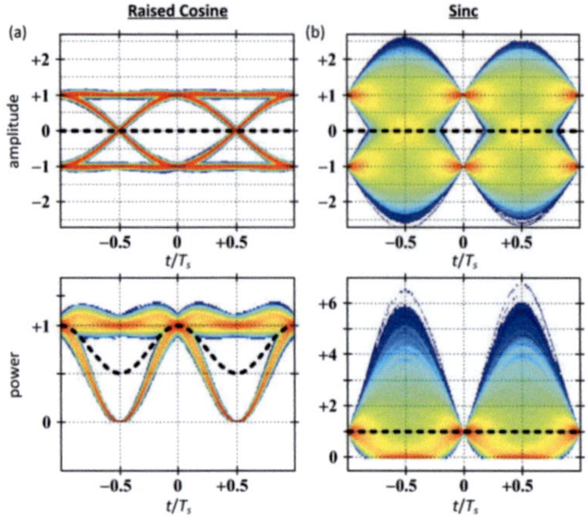

Fig. 2.25 Color-coded eye-diagrams and mean values (black dashed lines) as a function of time (T_s: symbol period). The plots are obtained from amplitudes (upper row) and powers (lower row) of binary phase shift keyed (BPSK) signals with (a) a sinc-like pulse shape with raised-cosine spectrum and (b) a sinc-pulse shape. While the mean power of signals with "raised-cosine" pulses (a) does reveal timing information, the signals comprising sinc-shaped pulses (b) produce large overshoots leading to a constant mean power for all t.

In a next step, we square the amplitudes [85] [86], so that power envelopes are obtained, see Fig. 2.25, lower row. To find the ideal sampling position with the maximum eye opening we now take the mean of the power envelope. It can be seen that in the case of a "raised-cosine" pulse shape, the mean power (black dashed line) is at its maximum at the ideal sampling position. Unfortunately, this simple operation does not help to find the ideal sampling time for the sinc-shaped signal, because here the mean power is constant for all possible sampling instances. This will be theoretically discussed in Section 2.4.6.2 where we show scenarios in which the mean power algorithm (MPA) performs well, and where we investigate its limitations.

In Section 2.4.6.3 we will then show how introducing a nonlinear operation [86] [87] prior to the averaging process will allow finding the ideal sampling time and carrier frequency even for sinc-shaped signals. It will be demonstrated that a variety of signals can be handled, while keeping the computational effort reasonably low.

2.4.6.2 The mean power algorithm for timing and frequency recovery

For deriving efficient methods for successfully recovering timing and carrier information from arbitrary Nyquist and OFDM signals, we now take a theoretical approach. In this section, we investigate to what extend the mean power of a signal can be used to obtain information with respect to carrier frequency offset and timing. First, sinc-shaped pulses/spectra (Nyquist/OFDM) are investigated, and then rect-shaped spectra/pulses (Nyquist/OFDM) will be discussed.

Nyquist timing and OFDM frequency recovery: Generally, a sinc-like Nyquist multi-carrier signal $s_{\mathrm{Nyquist}}(t)$ in the time domain, or an OFDM signal $s_{\mathrm{OFDM}}(f)$ in the frequency domain is defined by [79]

$$\begin{aligned} s_{\mathrm{Nyquist}}(t) &= \sum_{i=0}^{N-1}\sum_{k=-\infty}^{+\infty} c_{ik} \exp\left(+\mathrm{j}2\pi\frac{it}{T_s}\right)\operatorname{sincl}\left(\frac{t}{T_s}-k\right), \\ s_{\mathrm{OFDM}}(f) &= \frac{1}{F_s}\sum_{i=-\infty}^{+\infty}\sum_{k=0}^{N-1} c_{ik} \exp\left(-\mathrm{j}2\pi\frac{if}{F_s}\right)\operatorname{sinc}\left(\frac{f}{F_s}-k\right). \end{aligned} \tag{2.4.20}$$

The quantities c_{ik} are the complex modulation coefficients, T_s is the symbol period in the time domain, and F_s the symbol period in the frequency domain, i.e., the spacing of the OFDM subcarriers (SC). The sinc-like (sincl) and the sinc-functions are defined in the following:

$$\begin{aligned} \operatorname{sincl}\left(\frac{z}{Z}\right) &= \operatorname{sinc}\left(\frac{z}{Z}\right)\frac{\cos(\pi\beta z/Z)}{1-4\beta^2 z^2/Z^2}, \\ \operatorname{sinc}\left(\frac{z}{Z}\right) &= \begin{cases} 1 & \text{for } z=0 \\ \dfrac{\sin(\pi z/Z)}{\pi z/Z} & \text{else} \end{cases} \end{aligned} \tag{2.4.21}$$

The sincl-function has a raised-cosine Fourier transform [74] with roll-off factor β and a spectral support (bandwidth) of $(1+\beta)/Z$. Due to realization constraints, only a finite number of Q and R neighbors with respect to subscripts k and i contribute to the functions Eq. (2.4.20) for a given t or f. We also limit the number of symbols R with index i that will be considered. The expression in Equation (2.4.20) can then be formulated in a unified form with $z=t$, $Z=T_s$ (Nyquist) or $z=f$, $Z=F_s$ (OFDM, $\beta=0$)

$$s(z)=\sum_{k=-Q/2}^{+Q/2} c_k(z)\operatorname{sincl}\left(\frac{z}{Z}-k\right),\ c_k(z)=\sum_{i=-R/2}^{+R/2} c_{ik}\,\mathrm{e}^{\mathrm{j}2\pi\frac{iz}{Z}}. \tag{2.4.22}$$

The functions $c_k(z)$ are Fourier series and are represented by a finite number of $R+1$ terms subscripted with i due to the aforementioned realization constraints. For a single-carrier Nyquist signal or for a single-symbol OFDM spectrum ($R=0$, not a practical case!) this series reduces to one single term. Fig. 2.26 visualizes the signal Eq. (2.4.22) for $R=0$, $\beta=0$, and $Q=2$.

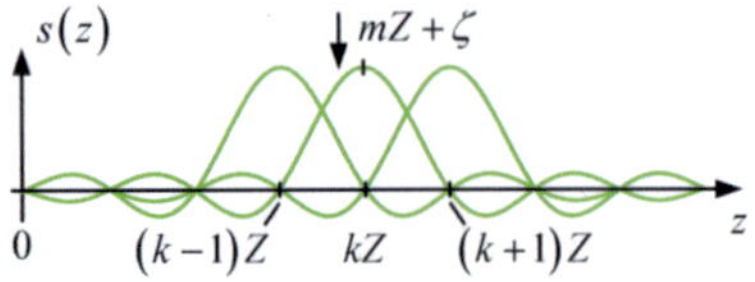

Fig. 2.26 Sinc-shaped pulse envelopes in time (Nyquist) or frequency domain (OFDM) as defined by the sum terms of Eq. (2.4.22) for $R = 0$ and $Q = 2$ neighbors. Pulses are separated by multiples k of the symbol period Z. The offset from the ideal sampling position mZ is named ζ.

In the following, we need to find the ideal sampling position for the reception of the signal described by Eq. (2.4.22). The ideal sampling position is located at the sincl-centers $z_m = m\,Z,\ m \in \mathbb{Z}$, see Fig. 2.26. Without loss of generality we set $m = 0$ and then define a so-called sampling offset ζ with respect to this ideal sampling position,

$$-0.5 < \zeta/Z \le 0.5 \tag{2.4.23}$$

From the intuitive introduction Fig. 2.25 we expect the optimum sampling position at $\zeta = 0$, i.e., at the position where the received signal power averaged over several samples is maximum. To show this, we now evaluate the ensemble average of the power $S(\zeta)$ of a Nyquist signal, or of an OFDM spectrum according to (Eq. (2.4.22) with $R = 0$ for a single carrier/symbol)

$$\begin{aligned} S(\zeta) = |s(\zeta)|^2 &= \sum_{k,k'=-Q/2}^{Q/2} c_k c_{k'}^* \operatorname{sincl}\left(\frac{\zeta}{Z} - k\right) \operatorname{sincl}\left(\frac{\zeta}{Z} - k'\right), \\ \overline{S(\zeta)} &= \sum_{k,k'=-Q/2}^{Q/2} \overline{c_k c_{k'}^*} \operatorname{sincl}\left(\frac{\zeta}{Z} - k\right) \operatorname{sincl}\left(\frac{\zeta}{Z} - k'\right) \\ &\approx \overline{|c_k|^2} \sum_{k=-Q/2}^{Q/2} \operatorname{sincl}^2\left(\frac{\zeta}{Z} - k\right). \end{aligned} \tag{2.4.24}$$

Because $R = 0$ is assumed in Eq. (2.4.22), the received modulation coefficients c_k are independent of the offset ζ. Furthermore, the lower row of Eq. (2.4.24) only gets contributions from the dominant sincl-contributions with $k = k'$ ($\overline{c_k c_{k'}^*} = 0$ for $k \ne k'$ and $\zeta \ne 0$ with statistically independent c_k and $c_{k'}$, which results from the symmetry property of commonly used constellation diagrams), so that the double sum reduces to a single sum. The quantity $\overline{|c_k|^2}$ stands for the mean power of all complex data coefficients c_k in Eq. (2.4.22). For sinc-functions ($\beta = 0$) and large numbers of signals/symbols $Q \to \infty$, the right-hand side of Eq. (2.4.24) reduces to $\overline{|c_k|^2}$. Thus, the mean power is constant and independent from the offset ζ. For $\beta \ne 0$ and/or for sufficiently small Q, we find that computing Eq. (2.4.24) provides a maximum mean power at the optimum sampling position $\zeta = 0$. Thus, in general, the optimum sampling position can be found by averaging the power over a small number of signals/symbols leading to the mean power algorithm (MPA). For visualization purposes, we plot the mean power according to Eq. (2.4.24) for different Q and β, see Fig. 2.27.

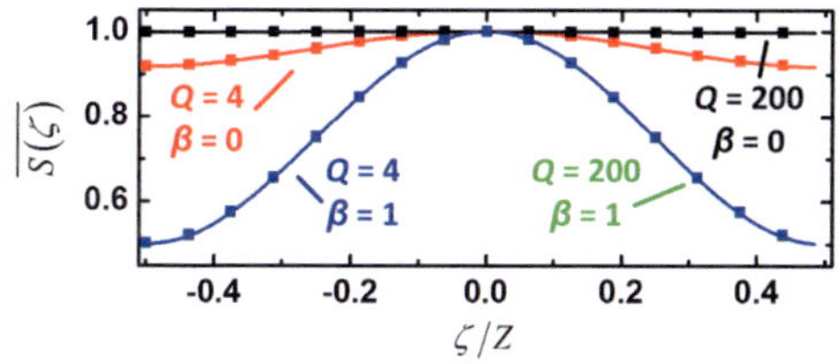

Fig. 2.27 For finding the ideal Nyquist sampling time or OFDM carrier frequency, the ensemble average of the power $S(\zeta)$ has been calculated (lines), simulated (solid squares), and plotted as a function of the sampling offset ζ normalized to the symbol period Z. The ideal sampling time or carrier frequency is given by $\zeta = 0$. Sinc-shaped functions with low roll-off factors β close to $\beta = 0$ only have a maximum at $\zeta = 0$ if a small number Q of neighbor symbols/spectra is included in the averaging process. For sincl-functions (having a raised-cosine spectrum) with β close to 1, the number of neighbor symbols Q is less important, because sincl-functions with larger β in Eq. (2.4.21) decay much faster than sinc-functions with $\beta = 0$.

To further support our findings, we additionally performed numerical simulations for random 64-ary quadrature amplitude modulated (64QAM) data. The analytically obtained curves (lines) and the results from simulations (squares) coincide. These general results can now be used to perform timing or carrier frequency recovery.

Timing recovery for Nyquist signals ($z = t$, $Z = T_s$, $\zeta = \tau$): The maximum of $\overline{S(\tau)}$ for a single-carrier Nyquist signal in Fig. 2.27 defines the correct sampling time $\tau = 0$ for the chosen pulse m. For sinc-pulses ($\beta = 0$), the number of neighbors Q must be sufficiently small for a well pronounced maximum. For sincl-pulses (having a raised-cosine spectrum with $\beta \neq 0$) even larger numbers of Q can be used.

Carrier frequency recovery for OFDM signals ($z = f$, $Z = F_s$, $\zeta = \nu$): The maximum of $\overline{S(\nu)}$ in Fig. 3 for a single-symbol OFDM signal defines the correct "sampling" frequency $\nu = 0$ for the chosen subcarrier m. Because ICI-free OFDM reception calls for orthogonal SC spectra (sinc-shaped envelope, $\beta = 0$), the number of neighbors Q must be sufficiently small, and only a few SCs can be included to find a well pronounced maximum in Fig. 2.27.

OFDM timing and Nyquist frequency recovery: Complementary to Eq. (2.4.20), we now formulate a rect-shaped OFDM baseband signal $s(t)$ in the time domain, or a sinc-shaped Nyquist multi-carrier signal with a rect-shaped spectrum $s(f)$ in the frequency domain [79],

$$s_{\text{OFDM}}(t) = \sum_{i=-\infty}^{+\infty} \sum_{k=0}^{N-1} c_{ik} \exp\left(+\text{j}2\pi \frac{kt}{T_s}\right) \text{rect}\left(\frac{t}{T_s} - i\right),$$
$$s_{\text{Nyquist}}(f) = \frac{1}{F_s} \sum_{i=0}^{N-1} \sum_{k=-\infty}^{+\infty} c_{ik} \exp\left(-\text{j}2\pi \frac{kf}{F_s}\right) \text{rect}\left(\frac{f}{F_s} - i\right). \tag{2.4.25}$$

The quantities c_{ik} are again the complex modulation coefficients, T_s is the OFDM symbol duration, and F_s the symbol "duration" in the frequency domain, i.e., the width of the Nyquist bands. The rect-function is defined by:

$$\text{rect}\left(\frac{z}{Z}\right) = \begin{cases} 1 & \text{for } |z| < Z/2 \\ 0 & \text{for } |z| > Z/2 \end{cases} \tag{2.4.26}$$

Again, only a finite number of R and Q neighbors with respect to subscripts i and k are considered within the functions Eq. (2.4.25) for a given t or f. The signals described by Eq. (2.4.25) can then be formulated in a unified form with $z = t$, $Z = T_s$ (OFDM) or $z = f$, $Z = F_s$ (Nyquist, $\beta = 0$) by

$$s(z) = \sum_{i=-R/2}^{+R/2} c_i(z) \text{rect}\left(\frac{z}{Z} - i\right), \quad c_i(z) = \sum_{k=-Q/2}^{+Q/2} c_{ik} \exp\left(\text{j}2\pi \frac{kz}{Z}\right). \tag{2.4.27}$$

The Fourier series $c_i(z)$ are again represented by a finite number of $Q + 1$ terms subscripted with k due to realization constraints. Fig. 2.28 visualizes the signal Eq. (2.4.27) for a single-symbol OFDM signal or for a single-carrier Nyquist spectrum ($R = 0$ and $Q = 2$). In this figure the window width is equal to the symbol width Z in time (OFDM) and frequency (Nyquist).

In the following we show that the ideal time (OFDM) or frequency (Nyquist) receiving window leads to a maximum when averaging the mean signal powers within each window. As before, the goal is to find either the ideal timing for the OFDM time window or the ideal center frequency for the respective Nyquist frequency window. In Fig. 2.28 this ideal windows would be centered at $z_m = m\,Z,\ m \in \mathbb{Z}$. Again, without loss of generality, we set $m = 0$.

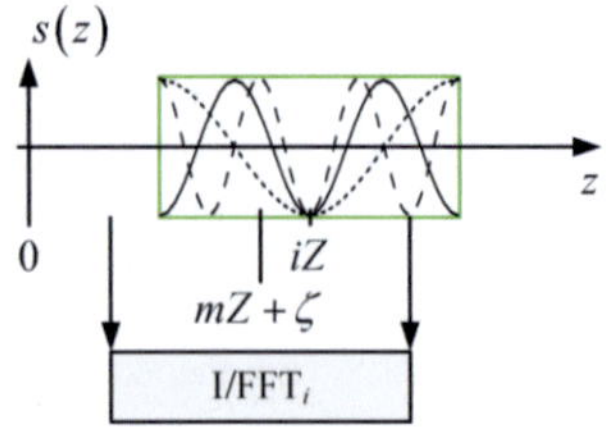

Fig. 2.28 Signal with a rect-shaped pulse envelope (green) showing either an OFDM signal in time domain or a Nyquist sinc-pulse signal in frequency domain (Eq. (2.4.25) for $R = 0$). The black curves show three superimposed "neighbors" (i.e., for $Q = 2$). The coefficients $c_{ik} = c_{0k}$ can be extracted by applying an FFT (for OFDM) or an IFFT (for sinc-pulses). An offset from the ideal I/FFT-window position is defined as ζ.

We now would like to find out to what extent the ideal timing or frequency at $\zeta = 0$ can be found by searching for the maximum in the ensemble average of the signal's mean power $S(\zeta)$ inside the receiving window, i.e., by evaluating

$$S(\zeta) = \frac{1}{Z} \int_{\zeta - Z/2}^{\zeta + Z/2} |s(z)|^2 \, \mathrm{d}z,$$

$$\overline{S(\zeta)} = \frac{1}{Z} \sum_{i,i'=-R/2}^{+R/2} \overline{\int_{\zeta - Z/2}^{\zeta + Z/2} c_i(z) c_{i'}^*(z) \operatorname{rect}\left(\frac{z}{Z} - i\right) \operatorname{rect}\left(\frac{z}{Z} - i'\right) \mathrm{d}z} \tag{2.4.28}$$

$$= \frac{1}{Z} \sum_{i=-R/2}^{+R/2} \overline{\int_{\zeta - Z/2}^{\zeta + Z/2} |c_i(z)|^2 \operatorname{rect}\left(\frac{z}{Z} - i\right) \mathrm{d}z}.$$

The rect-shaped symbols for an arbitrary R are non-overlapping, therefore the double sum over i, i' reduces to a single sum. If the receiving rect-window with width Z (integral boundaries) coincides with a rect-shaped symbol, we may ignore the rect-function and consequently the sum over i in Eq. (2.4.28). This is also true if the receiving rect-window overlaps with two neighboring symbols (no gap between symbols!), as far as the ensemble average is concerned. For an arbitrarily chosen i we then find by substituting $c_i(z)$ from Eq. (2.4.27) and by splitting $c_i(z)$ in a z-dependent and a z-independent part,

$$\overline{S(\zeta)} = \sum_{k,k'} \overline{c'_{ik} c'^*_{ik'}} \frac{1}{Z} \int_{\zeta - Z/2}^{\zeta + Z/2} \exp\left(\mathrm{j} 2\pi \frac{(k - k')z}{Z}\right) \mathrm{d}z = \sum_k \overline{|c'_{ik}|^2} \tag{2.4.29}$$

Due to inter-symbol interference (ISI), the received coefficients $c'_{ik} \neq c_{ik}$ belong to two different frames and do not correspond to the original data c_{ik}. The orthogonality integral in Eq. (2.4.29) reduces the double sum over k, k' to a single sum over k, and because the ensemble average of the modulus-squared coefficients $\overline{|c'_{ik}|^2} = \overline{|c_{ik}|^2}$ is same with and without ISI, Eq. (2.4.29) simplifies greatly. As a result, from measuring the average power in a window positioned arbitrarily at ζ, we cannot derive any synchronization information.

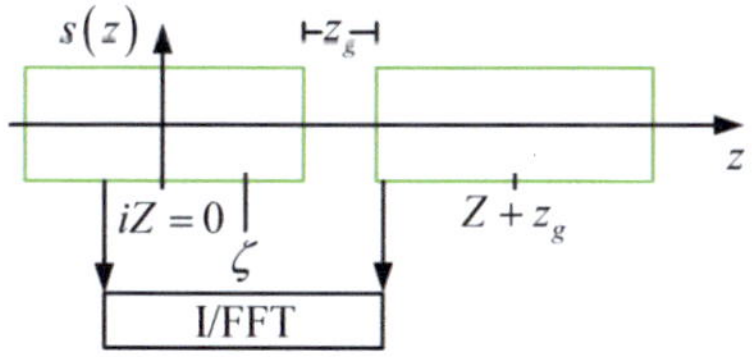

Fig. 2.29 Rectangular symbols and average power in receiving window as a function of the offset ζ. Signal comprising rect-shaped pulse envelopes (green) either in the time domain (OFDM) or in the frequency domain (Nyquist) separated by a guard interval z_g. The coefficients c'_{ik} are extracted within a rectangular window with an I/FFT.

The situation changes, if next neighbors are excluded, $R = 0$. In this case, the power as given by Eq. (2.4.29) drops if the receiving window is positioned wrongly,

$$\overline{S(\zeta)} = \sum_k \overline{\left|c'_{ik}\right|^2} \left(1 - \frac{|\zeta|}{Z}\right) \text{ for } R = 0 \text{ (no neighbors).} \tag{2.4.30}$$

Under this assumption, a peak in the detected triangular-shaped average power $\overline{S(\zeta)}$ of Eq. (2.4.30) indicates the optimum window position $\zeta = 0$.

Rather than excluding all neighbors, a small guard interval z_g in time (OFDM) or frequency domain (Nyquist) could be introduced, see Fig. 2.29. In this case, a peak in the mean power Eq. (2.4.30) indicates the optimum position for the chosen window m, even if neighbors are present, $R \neq 0$. To verify this, we compare theory (lines) to simulations (filled squares) in Fig. 2.30 and find that both agree well. These general results can now be used to perform timing or carrier frequency recovery.

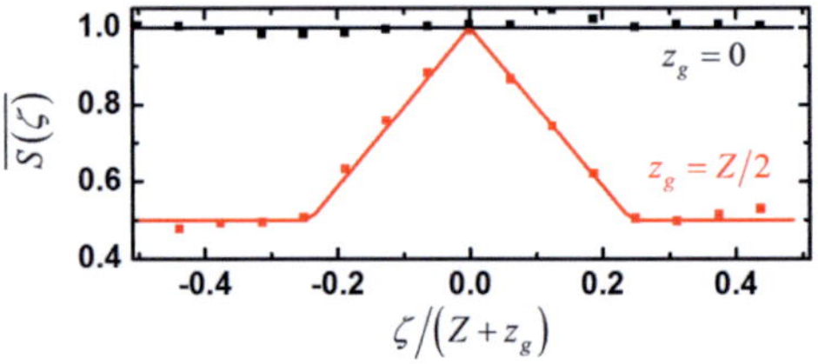

Fig. 2.30 The ideal OFDM time or the ideal Nyquist frequency window position, respectively, may be found by calculating the ensemble average of the power $S(\zeta)$ within the window as a function of the time/frequency offset ζ. Signals with a guard interval in time or frequency, i.e. with $z_g > 0$, have a distinct maximum exactly at the ideal timing or carrier frequency $\zeta = 0$. Therefore, time window or carrier frequency offsets can only be found for signals with a guard interval. The solid lines show the analytically obtained (Eq. (2.4.29) and (2.4.30)) results and the solid squares show results obtained by simulations.

Timing recovery for OFDM signals ($z = t$, $Z = T_s$, $\zeta = \tau$, $z_g = t_g$): For a single rectangular OFDM symbol ($R = 0$, not a practical case!), or if a guard interval t_g between adjacent OFDM symbols exists, the maximum of $\overline{S(\tau)}$ indicates the optimum FFT window position $\tau = 0$. Here, the samples contributing to $\overline{S(\tau)}$ are to be taken from the output of the FFT.

Carrier frequency recovery for Nyquist signals ($z = f$, $Z = F_s$, $\zeta = \nu$, $z_g = f_g$): For a single-carrier Nyquist signal ($R = 0$), or if a spectral guard band f_g exists, the maximum of $\overline{S(\nu)}$ indicates the optimum IFFT window position $\nu = 0$. In this case, the samples contributing to $\overline{S(\nu)}$ are to be taken from the output of the IFFT.

2.4.6.3 Mean modulus algorithm (MMA) and other nonlinear averages for timing and frequency recovery

Evaluation of the MPA $\overline{S(\zeta)}$ for finding the optimum recovery point $\zeta = 0$ does only yield the correct results if certain (sometimes unpractical) conditions are met. More precisely, and with the nomenclature of the previous section, the existence of a distinct maximum requires either sinc-like pulses with raised-cosine spectrum and a certain excess bandwidth $\beta > 0$, or a sufficiently small number of neighbors Q for sinc-shaped pulses/spectra as described by Eq. (2.4.24). In analogy, rect-shaped pulses/spectra such as described by Eq. (2.4.28) require a temporal or spectral guard interval z_g (or no next neighbor, $R = 0$).

However, in all cases, where evaluation of the mean power fails, one may apply a non-linear function $f(x)$ [86] to the received samples prior to averaging over a number of L measured values of $S(\zeta)$ in either time (Nyquist) or frequency domain (OFDM),

$$\overline{f\left(S(\zeta)\right)} = \frac{1}{L}\sum_{l=1}^{L} f\left(S(\zeta)\right)\Big|_l \qquad (2.4.31)$$

In Eq. (2.4.31), $f(x)$ can be, e.g., a square root function, a higher order polynomial, an exponential, a logarithm, or even a sinusoidal. Among all these possibilities, using a square root operation (i.e., an absolute-value rectifier) is very convenient as this means averaging over the signals' moduli. This leads us to an algorithm that we will subsequently refer to as the "mean modulus algorithm" (MMA). Computing the modulus and phase of a complex number in Cartesian coordinates is efficiently realized by the so-called CORDIC algorithm [92], where computationally expensive multipliers can be avoided. Furthermore, a transformation to polar coordinates is often employed anyhow when compensating the carrier phase offset, and hence does not consume any additional resources.

In conclusion, the two first moments defined by

$$\overline{S(\zeta)} = \frac{1}{L}\sum_{l=1}^{L} S(\zeta)\Big|_l \qquad (\text{MPA}) \qquad (2.4.32)$$

$$\overline{\sqrt{S(\zeta)}} = \frac{1}{L}\sum_{l=1}^{L} \sqrt{S(\zeta)}\Big|_l \qquad (\text{MMA}) \qquad (2.4.33)$$

can serve as cost (or objective) functions to efficiently recover timing and frequency information with little computational effort. As shown before, the MPA described by Eq. (2.4.32) has a limited area of application. While the MMA, Eq. (2.4.33), requires only little additional effort, it can be used to recover timing and frequency information of arbitrary Nyquist and OFDM signals. The wider range of application of the MMA can be attributed to the higher degree of non-linearity when applying a square root operation.

For completeness' sake, we put the MPA and MMA algorithm into the context of other, commonly used algorithms. For instance, the constant modulus algorithm (CMA) [93] is frequently used for polarization demultiplexing. In [79] and [43], it has been shown to also work for recovering frequency and timing information. The CMA cost function is defined by

$$\sigma^2_{\sqrt{S(\zeta)}} = \overline{\left(\sqrt{S(\zeta)} - \overline{\sqrt{S(\zeta)}}\right)^2} = \overline{S(\zeta)} - \overline{\sqrt{S(\zeta)}}^{\,2} \qquad (\text{CMA}) \qquad (2.4.34)$$

By rewriting the CMA cost function as per the second term in Eq. (2.4.34) it becomes obvious, that the CMA comprises the elements of the MPA and MMA algorithm as introduced in the previous section. The operation of the CMA is well understood for PSK signals, where it can be shown that the variance Eq. (2.4.34) is zero at the ideal sampling time $\zeta = 0$. This is valid independent of the pulse shape discussed previously. As an example, for $\zeta = 0$ the sum in Eq. (2.4.24) reduces to one term, $\sqrt{S(0)} = |c_0|$, or to $\sqrt{S(0)} = \sqrt{\sum_{k=-Q/2}^{+Q/2} |c_{0k}|^2}$ in the case of Eq. (2.4.28). Then the variance Eq. (2.4.34) becomes zero at $\zeta = 0$ and is larger in any other case. For all other QAM formats it is plausible to expect a minimum of the variance at $\zeta = 0$.

Instead of computing the variance of $\sqrt{S(\zeta)}$, one could alternatively evaluate the variance of the signal powers leading to the constant power algorithm (CPA) according to

$$\sigma^2_{S(\zeta)} = \overline{\left(S(\zeta) - \overline{S(\zeta)}\right)^2} = \overline{S^2(\zeta)} - \overline{S(\zeta)}^2 \qquad \text{(CPA)} \qquad (2.4.35)$$

If Eq. (2.4.35) is used as a cost function instead of the one for the CMA, square root operations can be avoided at the price of additional square operations. However, the general shape of the functions obtained from Eq. (2.4.34) and (2.4.35) is the same as will be shown later on.

The CMA and CPA cost functions Eq. (2.4.34) and (2.4.35) are combinations of the MMA and MPA algorithms given by Eq. (2.4.32) and (2.4.33). And indeed, the functions have been tested and work well for signals with sinc-pulses (Nyquist) or sinc-spectra (OFDM) [79] [43]. However, we are able to show that in most instances it is sufficient and more efficient to work with the MMA of Eq. (2.4.33) only. Since the MMA is a subset of the CMA, mathematical operations can be saved, hence reducing computational complexity and latency. In the following, we perform simulations for heavily noise loaded OFDM and Nyquist shaped signals and compare the convergence speed of the CMA and the MMA by determining the number of symbols required for reliable recovery of timing and frequency information. After that, a discussion of limitations and advantages of the exemplarily chosen cost functions Eq. (2.4.32)–(2.4.35) will be given along with experimental data.

2.4.6.4 Performance of MMA and CMA for noise-loaded signals

To find the number of symbols required to reliably recover timing and carrier information, we simulate QPSK and 16QAM modulated OFDM and Nyquist signals. Noise is added such that the resulting bit error ratio (BER) at the receiver is either 10^{-3} or 10^{-2}. This is close to state-of-the-art hard-decision and soft-decision forward error correction limits. For each BER and each number of received coefficients, timing and frequency information is determined for a 100 times.

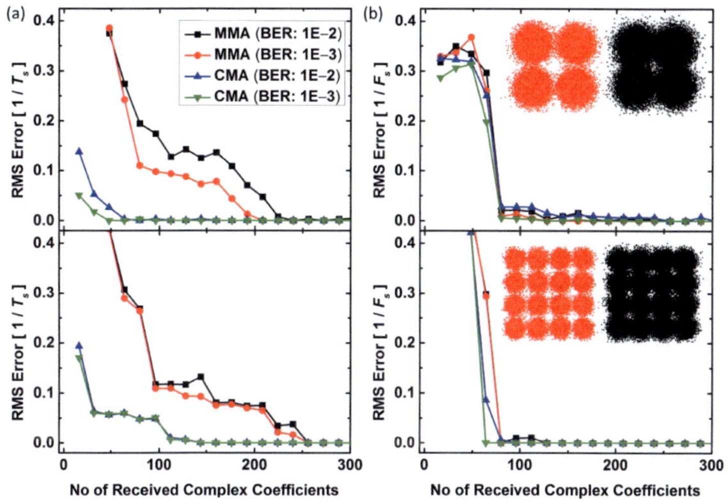

Fig. 2.31 Performance of MMA and the CMA for simulated QPSK (top) and 16QAM (bottom) OFDM signals with noise loading. (**a**) RMS of error made by (**a**) timing or (**b**) frequency recovery.

We then compute the root-mean-square (RMS) of the error made by the recovery algorithms normalized to either symbol period or bandwidth. Results for OFDM are shown in Fig. 2.31. In Fig. 2.31(a) we compare the performance of the MMA and the CMA when being used for timing recovery. It can be seen that the CMA requires less received complex coefficients to reliably determine the FFT window position at the Rx. However, the MMA can handle the increased amount of required coefficients within only two additional clock cycles (assuming a binary adder tree) leading to a net latency and complexity comparable to the CMA where differences between mean modulus and individual moduli have to be summed. When used for OFDM frequency recovery, see Fig. 2.31(b), both algorithms require the same amount of coefficients.

Results for QPSK and 16QAM Nyquist sinc-pulses are depicted in Fig. 2.32. Again timing recovery, see Fig. 2.32(a), and frequency recovery, see Fig. 2.32(b), is performed with either the CMA or the MMA.

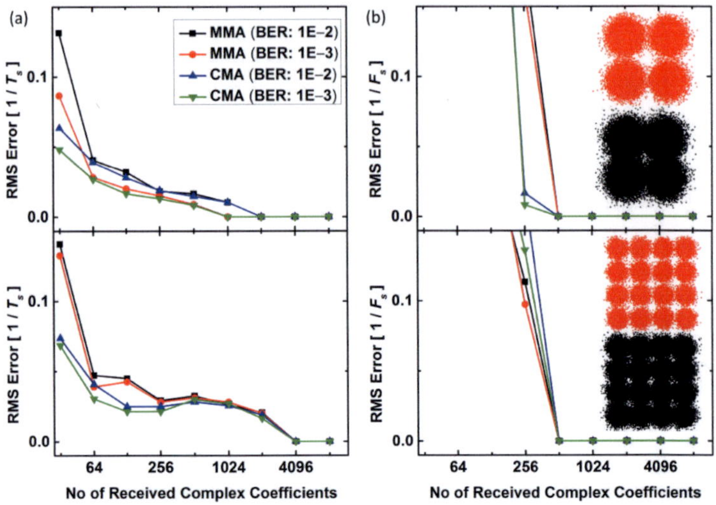

Fig. 2.32 Performance of the MMA and the CMA for simulated QPSK (top) and 16QAM (bottom) Nyquist signals. (**a**) RMS of error made by (**a**) timing or (**b**) frequency recovery.

From these results we conclude that the proposed MMA and the CMA perform equally well, leaving the MMA with the advantage of less computational complexity except for OFDM timing synchronization where the complexity is virtually same for both algorithms.

2.4.6.5 Experimental Setup

The experimental setup is depicted in Fig. 2.33. A software-defined transmitter (Tx) [73] acts as an arbitrary waveform generator (AWG), i.e., two synchronized field programmable gate arrays (FPGA) store pre-computed waveforms. With these waveforms the FPGAs drive two Micram digital-to-analog converters (DAC) with a resolution of 6 bit operating at a sampling rate of 28 GSa/s. The in-phase (I) and quadrature (Q) component of the complex Nyquist or OFDM waveforms are encoded on an external cavity laser (ECL) with a center wavelength of 1550 nm. To do so, we use an I/Q-modulator consisting of nested $LiNbO_3$ Mach-Zehnder modulators and a $\pi / 2$ phase shifter. An erbium-doped fiber amplifier (EDFA) amplifies the optical signals. A 1 nm optical band-pass filter removes out-of-band EDFA noise. Finally, an Agilent optical modulation analyzer (OMA) coherently receives the signals using a free-running local oscillator (LO). The in-built real-time oscilloscope samples with a rate of 80 GSa/s. All signal processing is done offline.

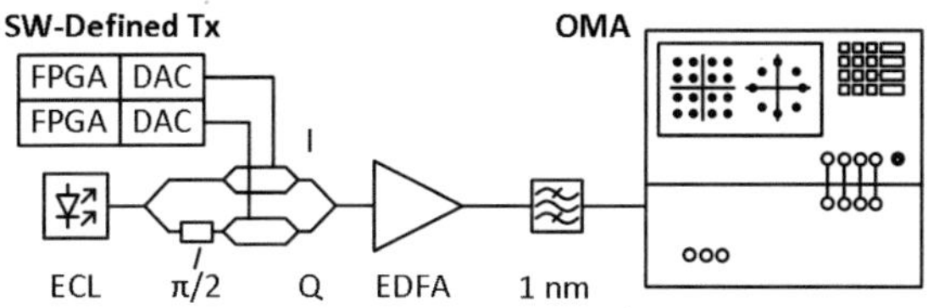

Fig. 2.33 Experimental setup for transmitting Nyquist or OFDM signals. Two synchronized field programmable gate arrays (FPGA) store pre-computed waveforms and drive attached digital-to-analog converters (DAC). The generated complex waveform is then encoded on an external cavity laser (ECL) with an optical I/Q-modulator. After amplification with an erbium doped fiber amplifier (EDFA) the signal passes a 1 nm-wide bandpass, is coherently received by an optical modulation analyzer (OMA), and processed offline.

2.4.6.6 Experimental Results for Carrier Frequency and Timing Recovery

To verify the proposed frequency and timing recovery mechanisms experimentally, we use the setup depicted in Fig. 2.33. We transmit Nyquist signals having different roll-off factors β, and ideally rect-shaped OFDM signals. In both cases we employ the formats QPSK, 16QAM (only OFDM) and 64QAM. We evaluate the various cost functions Eq. (2.4.32)–(2.4.35) and determine either a temporal sampling offset $\zeta = \tau$ or a carrier frequency offset $\zeta = \nu$.

Orthogonal Frequency Division Multiplexing (OFDM): We first evaluate the frequency recovery mechanism for OFDM signals comprising 128 SCs, each modulated with QPSK (black), 16QAM (blue), or 64QAM (red).

Carrier frequency recovery ($z = f$, $Z = F_s$, $\zeta = \nu$, $\tau = 0$): We vary the frequency of the LO and extract the modulation coefficients c'_{ik} with an FFT, the window of which is positioned at the ideal synchronization point $\tau = 0$. With the resulting coefficients c'_{ik}, we evaluate the MPA, the MMA, the CMA, and the CPA algorithms according to Eq. (2.4.32)–(2.4.35) as a function of the frequency offset ν normalized to the subcarrier spacing F_s (symbol rate). The outcome is depicted in Fig. 2.34. All curves are normalized to their maximum values.

With all methods except for the MPA, the zero offset $\nu = 0$ between the signal's carrier frequency and the LO frequency is found at extremal points, independent of the modulation format, see Fig. 2.34.

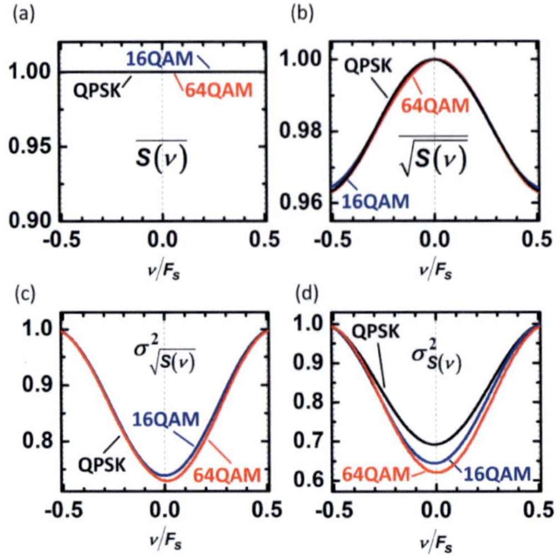

Fig. 2.34 Experiment showing the performance of four cost functions for OFDM carrier frequency recovery for either QPSK (black), 16QAM (blue), or 64QAM (red) modulated SCs. (**a**) The MPA does not provide an extremum and thus cannot be used to recover the frequency. (**b**) The mean modulus algorithm (MMA) of the extracted coefficients c_{ik} yields a maximum at the frequency offset $\nu = 0$. (**c**) The CMA and (**d**) the CPA provide a minimum for $\nu = 0$.

Timing recovery ($z = t$, $Z = T_s$, $\nu = 0$, $\zeta = \tau$): As a next step we investigate the previously discussed algorithms and the respective cost functions for recovering the timing of OFDM signals. In analogy to the experiment discussed above, we extract the modulation coefficients c'_{ik} for each symbol. This time, however, we fix the frequency offset ν at zero and vary the temporal offset τ of the FFT window. The results are shown in Fig. 2.35. For Fig. 2.35(a) only, computing the MPA of simulated OFDM signals with a temporal guard interval τ_g between adjacent symbols gives a distinct maximum where the temporal offset τ equals zero. These results agree well with the predictions made by Eq. (2.4.30). In Fig. 2.35(b)–(d) the temporal window position of measured OFDM signals is evaluated using either the MMA (Fig. 2.35(b), Eq. (2.4.33)), the CMA (Fig. 2.35(d), Eq. (2.4.34)) or the CPA (Fig. 2.35(d), Eq. (2.4.35)). All of these methods can be used as part of a feedback loop that provides a continuous OFDM symbol synchronization.

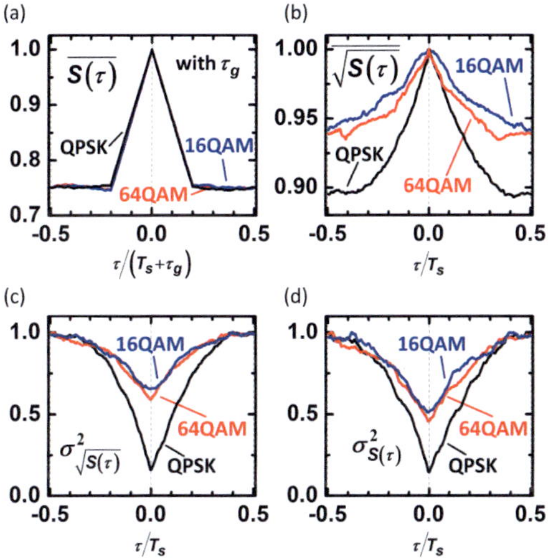

Fig. 2.35 Demonstration of the performance of four cost functions for OFDM timing recovery. (**a**) MPA cost function for various offset times for simulated OFDM symbols with a temporal guard interval τ_g in-between symbols. (**b**) – (**d**) show the result of applying the cost functions to experimentally obtained OFDM signals using either (b) the MMA, (c) the CMA or (d) the CPA algorithms. These algorithms provide an extremum at the ideal time even if $\tau_g = 0$.

Nyquist Signaling: We now generate QPSK and 64QAM Nyquist signals with different roll-off factors β and investigate the proposed techniques for frequency and timing recovery.

Carrier frequency recovery ($z = f$, $Z = F_s$, $\zeta = \nu$, $\tau = 0$)*:* For frequency recovery, a number of three simultaneous Nyquist signals ($\beta = 0$) centered at different carrier frequencies are transmitted. Only the middle channel is considered while the neighboring Nyquist channels are removed by rectangular digital filters having a pass-band equal to the channel bandwidth. The outcome of the proposed algorithms and cost functions as a function of frequency offset ν is depicted in Fig. 2.36. For evaluation of the MPA, see Fig. 2.36(a), we perform simulations and introduce a spectral guard interval ν_g in-between the Nyquist channels. Again, the results agree with the predictions made by Eq. (2.4.30), and a distinct maximum can be identified where $\nu = 0$. For the remaining plots the techniques even work without guard intervals. Experiments are again performed with the setup in Fig. 2.33. It can be seen that the evaluation of the MMA yields a maximum at $\nu = 0$, see Fig. 2.36(b) and Eq. (2.4.33). Furthermore, the CMA, Fig. 2.36(c), and the CPA, Fig. 2.36(d), both show a minimum if the LO frequency is synchronized to the carrier frequency of the Nyquist signal.

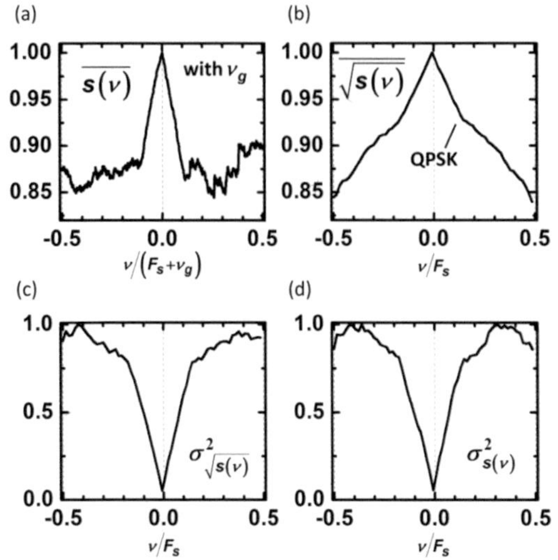

Fig. 2.36 The performance of four cost functions for Nyquist carrier frequency recovery. (**a**) MPA cost function as a function of frequency offsets applied to a simulated Nyquist signal with spectral guard interval ν_g in-between channels. (**b**) – (**d**) Results of cost functions obtained by applying the operations on experimentally received Nyquist signals without a guard interval using either (b) the MMA, (c) the CMA or (d) the CPA algorithms.

Timing recovery ($z = t$, $Z = T_s$, $\nu = 0$, $\zeta = \tau$): For evaluation of the timing recovery, we transmit single-channel Nyquist signals using "raised-cosine" pulse-shapes Eq. (2.4.21) with different roll-off factors β, modulated with either QPSK or 64QAM. The results are shown in Fig. 2.37. In each of the subfigures Fig. 2.37(a) and (b) we depict the mean power (MPA, upper left), the mean modulus (MMA, upper right), the variance of the modulus (CMA, lower left), and the variance of the modulus squared (CPA, lower right) as a function of the temporal offset τ from the ideal sampling positions. All cost functions can be used to find the proper sampling times for QPSK, see Fig. 2.37(a), and 64QAM signals, see Fig. 2.37(b). However, as predicted by Eq. (2.4.24), employing the MPA is not useful for finding the proper timing if signals with $\beta < 0.3$ are received. The curves obtained with Eq. (2.4.32) (MMA) show a pronounced maximum at $\tau = 0$ even for $\beta = 0$, and for large supports Q. For Nyquist signals with $\beta > 0.3$, both methods based on the variance (CMA and CPA) fail for high order M-ary QAM. In these scenarios the MPA or the MMA should be used. In Fig. 2.37(c) we depict measured constellation diagrams for QPSK and $\beta = 0$ (top) and for 64QAM and $\beta = 1$ (bottom). Sampling with $\tau = 0$ yields the red symbols, whereas sampling with $\tau \neq 0$ will result in the black inter-symbol transitions, i.e., sampling is done at the slope of the sinc-function. The discussed timing recovery for Nyquist signals is robust with respect to carrier frequency and carrier phase offsets since the phase information is not regarded. Furthermore, the method tolerates differential group delays >100 ps.

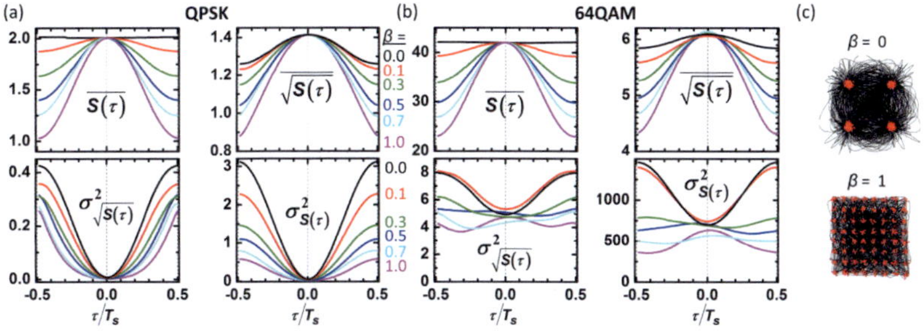

Fig. 2.37 Performance of four cost functions to recover the timing offset for Nyquist signals with different roll-off factors β. We evaluate the MPA (upper left), the MMA (upper right), the CMA (lower left), and the CPA (lower right) for (**a**) QPSK and (**b**) 64QAM encoded Nyquist signals and as a function of the temporal sampling offset τ. The MPA only yields a good measure for $\beta > 0.1$. Another limitation arises for Nyquist signals with large β where the CMA and CPA algorithms do not provide a distinct minimum for $\tau = 0$ (e.g. for 64QAM). (**c**) Constellation diagrams for QPSK and $\beta = 0$ (top) and for 64QAM and $\beta = 1$ (bottom). Sampling with $\tau = 0$ yields the red symbols whereas sampling with $\tau \neq 0$ results in the black inter-symbol transitions.

2.4.6.7 Carrier Phase Recovery

After frequency and timing recovery, we still need to compensate the phase offset φ between the signal carrier and the LO. To do so, the Viterbi-Viterbi algorithm [90] is typically employed. Unfortunately, the computational effort scales with the degree of phase modulation so that especially for high order M-ary QAM with μ different phases a significant amount of computational effort is required (the complex valued signal has to be taken to the power of μ). As a low-complexity alternative we suggest to apply the mean modulus algorithm (MMA) to the real and imaginary parts of the received symbols according to

$$\begin{aligned}
&\overline{\left|\Re\{s(\varphi)\}\right| + \left|\Im\{s(\varphi)\}\right|} = \frac{1}{M}\sum_{m=1}^{M}\left[\left|\Re\{c_m e^{j\varphi}\}\right| + \left|\Im\{c_m e^{j\varphi}\}\right|\right],\\
&c_m = |c_m| e^{j\theta_m},\\
&\Re\{c_m e^{j\varphi}\} = |c_m|\left[\sin(\theta_m)\sin(\varphi) - \cos(\theta_m)\cos(\varphi)\right],\\
&\Im\{c_m e^{j\varphi}\} = |c_m|\left[\sin(\theta_m)\cos(\varphi) + \cos(\theta_m)\sin(\varphi)\right].
\end{aligned} \tag{2.4.36}$$

The outcome predicted by Eq. (2.4.36) has been compared to simulations. The results are illustrated in Fig. 2.38. It can be seen that the MMA provides the ideal phase at the maximum. Fig. 2.38 also shows that Eq. (2.4.36) (solid lines) and simulations (squares) coincide. We evaluated QPSK, Fig. 2.38(a), 16QAM, Fig. 2.38(b), 32QAM, Fig. 2.38(c), and 64QAM data, see Fig. 2.38(d). Insets show constellation diagrams with different phase offsets φ. While the Viterbi-Viterbi algorithm requires a significant amount of multipliers, the algorithm according to Eq. (2.4.36) does not require any multipliers at all and is thus highly beneficial for implementations with emphasis on low computational effort. As with the Viterbi-Viterbi algorithm, there is a phase ambiguity as the received constellation could be rotated by multiples of $\pi / 2$ (for quadratic constellations) [90]. This has to be handled separately, e.g., by training symbols.

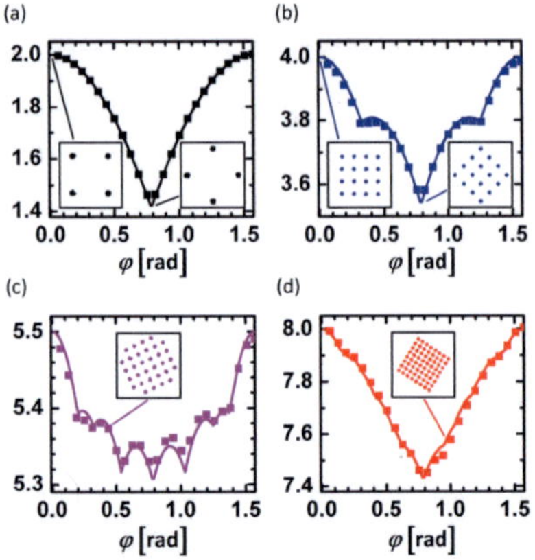

Fig. 2.38 Carrier phase estimation by applying the MMA to the real and imaginary values of the signal: Theory (lines, Eq. (2.4.36)) and simulations (squares) agree well. Typically, an offset φ between signal carrier phase and LO phase is observed. Constellation diagrams for different φ are shown as insets. **(a)** QPSK. **(b)** 16QAM. **(c)** 32QAM. **(d)** 64QAM.

2.4.6.8 Gardner Timing Recovery

The timing recovery after Gardner [85] can be used to synchronize the Rx clock to the Tx clock if signals with excess bandwidth between 40% and 100% are transmitted [85]. This can be, e.g., RC-signals with roll-off factor $0.4 \le \beta \le 1$ [74] whereas the timing of RC-signals with small spectral roll-off factor $\beta < 0.4$ cannot be recovered this way. The Gardner timing recovery uses two samples per symbol to identify the clock rate and phase through symbol transitions. The functionality is best understood when looking at an analog waveform $u(t)$ in the baseband with a symbol period T_s and carrying binary data $c = \pm 1$ as depicted in Fig. 2.39(a). Samples taken at the center of each symbol are indicated by black arrows. First, the analog waveform $u(t)$ is rectified with a square-law rectifier and $u^2(t)$ is obtained, see Fig. 2.39(b). In $u^2(t)$, a DC-component as well as a double frequency component (at the symbol frequency) can be identified. It should be further noted that, due to square-law rectification, the carrier phase need not be known and is hence independent of the timing recovery. A narrow-band analog filter with its passband at the symbol frequency (or symbol rate) could now be used to extract the clock from the rectified signal $u^2(t)$. In a digitial system the same approach could be used, which has been proposed in [88]. Here, this narrow-band filter is realized in the digital domain by employing a fast Fourier transform (FFT). The digitally rectified signal is fed to an FFT and the resulting amplitude spectrum is analyzed. A spectral line appears at the clock frequency and the clock phase can be extracted from that line. Also this so-called squared timing recovery is only applicable to signals with an excess bandwidth > 40%.

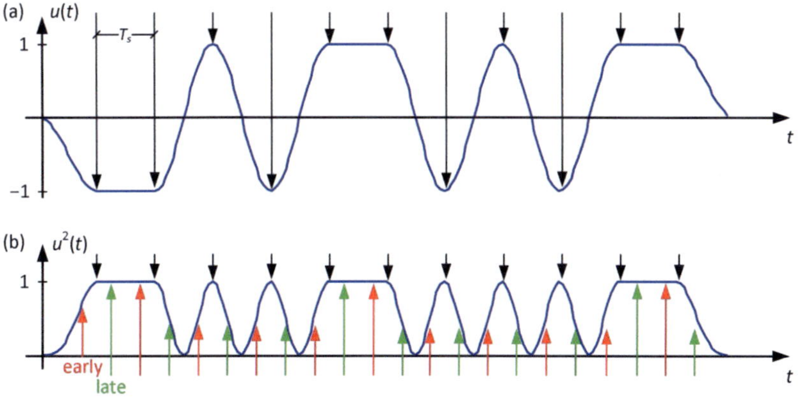

Fig. 2.39 Gardner timing recovery illustrated using a real signal $u(t)$ with symbol period T_s and with modulation coefficients $c = \pm 1$. (**a**) Sampling instances taken at the symbol centers are indicated by black arrows. (**b**) After rectification with a square-law rectifier, the signal $u^2(t)$ is obtained. Using two-fold oversampling, the difference of early taken samples (red arrows) and late taken samples (green arrows) equals zero, if these two samples (red and green) are centered at the correct sampling instances (black arrows) and if signals without noise and symmetric pulse-shape are assumed.

For the Gardner timing recovery [85], the timing error is to be determined rather than the clock itsels as is done in the squared timing recovery [88]. Therefore, it can be shown that at sample points that are $\pm 1/4$ off the symbol centers(see 'early' and green 'late' arrows in Fig. 2.39(b)), equal amplitudes are seen if a transition between adjacent symbols occurs. Therefore if the timing is correct, then the difference between the amplitude of the sample taken $-1/4$ off the symbol center and the amplitude of the sample taken $+1/4$ off the symbol center is zero. If there are no transitions between the early sample ($-1/4$ off, red) and the late sample ($+1/4$ off, green), then the difference is only zero if no noise is present. A modified algorithm that is more robust towards noise, uses a prefilter to remove sequences of symbols without transisition in between.

2.4.6.9 Carrier Recovery with the Vector Signal Analyzer Software

The vector signal analyzer (VSA) software provided by Agilent is frequently used throughout the presented experiments. The software achieves frequency offset compensation and carrier phase estimation in two steps. As an example a QPSK signal with a frequency offset and arbitrary carrier phase is depicted in Fig. 2.40. Due to the frequency offset the constellation diagram consisting of four points rotates and therefore forms a circle when observed over multiple symbol lengths, see Fig. 2.40(a).

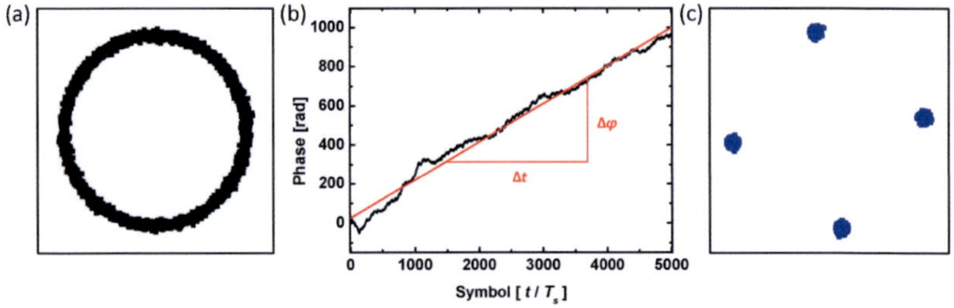

Fig. 2.40 Frequency offset and carrier phase estimation for a QPSK signal performed by the Agilent vector signal analyzer (VSA) software. (**a**) A frequency offset leads to a rotation of the constellation diagram. (**b**) The phase change over time reveals the frequency offset. Due to phase modulation of the signal, a linear regression determines the slope. (**c**) Frequency offset compensated signal that shows a constant phase offset within the observation period. The phase offset is compensated for by minimizing the error vector magnitude (EVM) of the received signal.

The phase change over time illustrated in Fig. 2.40(b) reveals the frequency offset. Since received signals usually have a phase modulation, the phase change over time deviates from a straight line as would be expected from a constant frequency offset. Assuming arbitrarily encoded data where each symbol in the constellation diagram is chosen with equal probability, linear regression provides the required information. To improve the frequency estimation one may remove the phase modulation by raising the signal to the power of M (with M being the number of possible phase states in the signal). This is known as the Viterbi-Viterbi recovery and will be explained in more detail in Section 2.4.6.10. Finally, the carrier phase has to be estimated. A constellation diagram after frequency offset compensation is shown in Fig. 2.40(c). For the carrier phase estimation it is required that the phase is constant within the observation time. The software then applies several phase shifts and determines the non-data-aided error vector magnitude (EVM). To do so, only the modulation format has to be made known to the software. This quality metric is further explained in the Appendix A.1. The optimum phase is found where the EVM is minimum. There are several simplifications that can be made, e.g. a phase variation of $\pi / 2$ only suffices to find the optimum position due to the symmetry of the constellation diagrams of square-shaped QAM formats.

2.4.6.10 Viterbi-Viterbi Carrier Recovery

Another way to estimate the carrier phase of a coherently received and sampled signal $u(n)$ is to use the Viterbi and Viterbi estimator [90]. For this estimation it is crucial to know the number of different phases that are present for a given modulation format. The QPSK format, e.g., produces four different points or symbols in the constellation diagram that have the same modulus (distance from the origin) but four different phases. In order find the phase offset $\Delta\varphi$ caused by the phase difference of transmitter laser φ_{Tx} and local oscillator φ_{Rx}, a received and M-ary phase modulated signal $u(n)$ is to be taken to the M-th power to remove the phase modulation. The phase offset $\Delta\varphi$ can then be found according to [59]

$$\Delta\varphi = \Phi\left\{\sum_{n=n-k}^{n+k} u^M\left(nT_s\right)\right\} \Big/ M, \tag{2.4.37}$$

where the phase information of the sum of $2k + 1$ samples each raised to a power of M is extracted with the $\Phi\{\cdot\}$-operator. The more $2k + 1$ samples taken into account, the more accurate the phase estimation gets. However, it is required that within the observation time $(2k + 1)T_s$, the laser induced phase $\Delta\varphi$ is constant. The phase estimation obtained from Eq. (2.4.37) varies between $-\pi / M$ and $+\pi / M$. If needed so-called unwrapping of the obtained phase may be performed according to the following relations:

$$\Delta\varphi_{\text{unwrapped}} = \Delta\varphi + \frac{2\pi}{M} f\left(\Delta\varphi\right) \tag{2.4.38}$$

$$f(x) = \begin{cases} +1 & x < -\pi/M \\ 0 & |x| \le +\pi/M \\ -1 & x > +\pi/M \end{cases} \tag{2.4.39}$$

With the estimated phase offset $\Delta\varphi$ obtained from Eq. (2.4.37), the phase can be corrected according to

$$u_{\mathrm{VV}}(n) = u(n)\exp\left(-\mathrm{j}\Delta\varphi + \pi/4\right). \tag{2.4.40}$$

An additional phase shift of $\pi / 4$ in Eq. (2.4.40) is applied as the Viterbi and Viterbi algorithm otherwise aligns the e.g. QPSK constellation diagram to the real and imaginary axis. In common standards the QPSK constellation diagram is turned by a 45° angle with respect to the two axes, see Fig. 2.41.

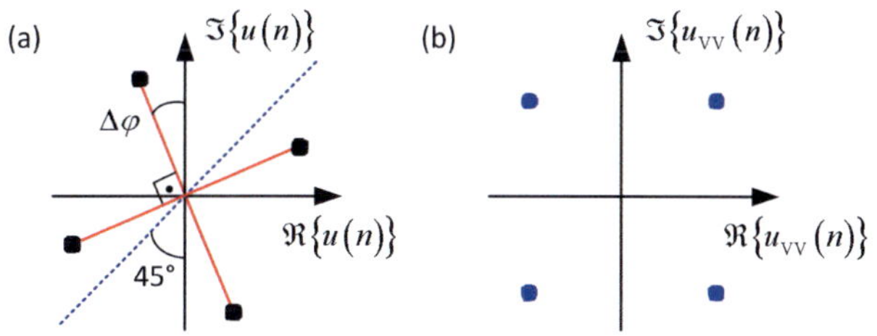

Fig. 2.41 Phase offset compensation for a QPSK signal using the Viterbi-Viterbi algorithm. (**a**) There is a phase offset $\Delta\varphi$ between carrier phase and the local oscillator (LO) phase. (**b**) The constellation diagram is aligned and $\Delta\varphi = 0$.

Unfortunately, there is a phase ambiguity as the QPSK constellation diagram along with all quadratic M-ary QAM formats is symmetric, i.e., there are four possible phases (each separated by 90°) that lead to an optimum constellation diagram. In order to deal with this ambiguity, training symbols or preambles help to determine which of the four solutions is correct. The problem can be avoided if differential phase encoding has been selected. With this technique, the phase information of the n-th symbol is encoded taking the phase information of the $(n-1)$-th as a reference [94]. However, differential encoding leads to a penalty doubling the BER.
Another drawback of the Viterbi and Viterbi estimator is the restricted capability to adapt to modulation formats with multiple amplitude levels such as M-ary QAM. Finally, the computation complexity of this phase estimator scales with the number of phases used by a given modulation format. When for QPSK the 4th power of $u(n)$ has to be computed, 8-PSK already requires the computation of a power of 8. This complexity increase inherently limits the application area of the Viterbi and Viterbi algorithm.

2.4.6.11 Data-Aided Recovery

When the previously described algorithms for carrier and clock recovery operate blindly, i.e., without knowledge of the transmitted data, there are also schemes that rely on training symbols or pilot tones. Data-aided synchronization is especially useful for bursty traffic where the synchronization process in the Rx needs to lock quickly and where no continuous locking can be achieved. In these scenarios also so-called feedforward recovery schemes are applied instead of the less resource hungry feedback schemes.

2.4.6.12 Schmidl-Cox Synchronisation for OFDM

The most popular synchronization algorithm for OFDM signals is the Schmidl-Cox synchronization. This algorithm is based on a correlation technique to find the start of an OFDM frame and therefore the proper FFT windows at the Rx [91]. To find a distinct maximum in the correlation, special training symbols are send, where first only the odd subcarriers (SC) are modulated with pseudo random data. In a second training symbol, both even and odd SCs carry pseudo random data. If the OFDM signals uses a cyclic extension [23] to account for finite rise and fall times of the electronic or to compensate for CD, the single maximum of the correlation becomes a plateau leaving some ambiguity with respect to the frame start. This can be avoided by using training symbols without this cyclic extension, while keeping the cyclic extension for the remaining OFDM symbols. A potential frequency offset of the LO with respect to the signal carrier does not affect the performance of the timing recovery, since this offset affects all training symbols equally, and the correlation still exhibits a maximum at the point of temporal synchronization. Therefore, after timing recovery, the phase offset between adjacent training symbols is used to determine the carrier frequency offset. This offset can be separated into a fraction-al offset (which is smaller than the carrier spacing) and an offset that is an integer multiple of the carrier spacing. Both frequency offsets can be determined by evaluation of the phase evolution within the training symbols.

2.4.7 Linear Equalization

Linear equalization is inevitable for almost every communication link as the frequency responses of built-in components such as amplifiers or filters are neither constant in amplitude nor constant in phase throughout the frequency range of interest. These linear equalizers mainly mitigate patterning effects as well as inter-symbol interference (ISI) that is caused by the frequency response of the overall transmission channel. Equalization can either be performed with analog components such as resistors, capacitors, and inductors, or it can be realized in the digital domain. In the latter case the signal is sampled at equidistant times and quantized so that the computation may be performed with a finite resolution. There are essentially two types of linear equalizers:

1. *Finite-duration impulse response (FIR) filter* where the impulse response $h(t)$ is represented by a finite number of filter coefficients h_r. Its structures contain feed-foward paths only. Therefore an FIR filter is inherently stable.
2. *Infinite-duration impulse response (IIR) filter* where the impulse response $h(t)$ is infinitely extended in time. In contrast to FIR filters, the IIR filter includes also feedback paths. Therefore special precaution must be taken to guarantee stability.

In this work, the focus lies on time discrete digital equalizers with a finite-duration impulse response (FIR). Further one can classify equalizers to be either static or adaptive. If the equalizer is static, the set of filter coefficients h does not change. For these equalizers the channel properties must be known precisely as they determine the performance of the static equalizer. Adaptive equalizers start from an initial set of coefficients $h(0)$ that are adjusted in the course of operation. Nevertheless a well-chosen initial set of coefficients can significantly enhance the convergence of these filters. The following sections give more details about time discrete linear FIR filters along with adaption techniques based on the least mean square (LMS) algorithm.

2.4.7.1 Time Discrete Linear Equalizers

Assuming that the appropriate sampling rate and a sufficient number of quantization levels have been selected, any linear filter operation can be performed by digital means, i.e., the filter output $y(n)$ is obtained from a discrete convolution of the sampled input $u(n)$ with the sampled filter response h represented by the filter coefficients h_r.

$$y(n) = \sum_{r=0}^{R} h_r u(n-r) = hu^T(n) \qquad (2.4.41)$$

Here, R denotes the filter order and $u^T(n)$ is the transposed input sample vector $u(n)$. To realize the filter output in Eq.(2.4.41), there are three types of time discrete FIR filters:

1. *Transversal filter*
2. *Lattice predictor*
3. *Systolic array*

The most common and straight forward FIR filter architecture is the transversal filter, also known as direct form I filter. The transversal filter comprises a series of R delay elements with unit delay z^{-1} of the Z-Transform, see Fig. 2.42. The delayed signal is tapped (thus the term filter taps) and weighed by the individual and possibly adaptive filter coefficients h_r. The sum of the tapped and weighed signal finally forms the filter output $y(n)$.

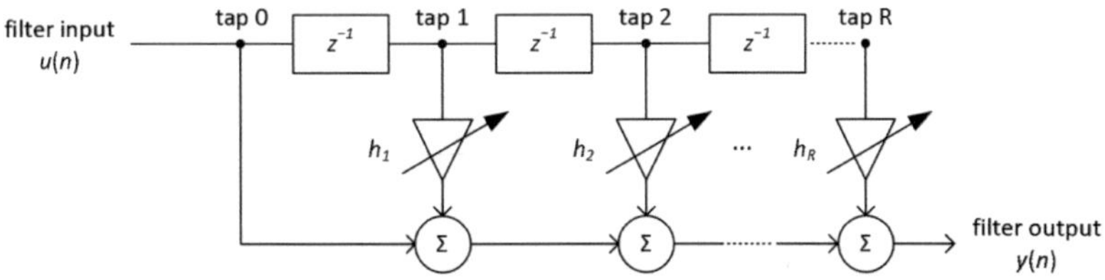

Fig. 2.42 Transversal finite-duration impulse response (FIR) filter. The filter of order R comprises R delay elements with unit delay z^{-1} and R possibly adaptive weights represented by the filter coefficients h_r. The tapped and weighed input signal $u(n)$ is summed to form the filter output $y(n)$.

2.4.7.2 Wiener Filters

One class of linear optimum time-discrete filters is referred to as Wiener filters [95]. These filters are optimum in a sense that they maximize the signal-to-noise ratio (SNR) of a noise corrupted signal. There are some practical restrictions that should be met when constructing a Wiener filter:

1. The filter is linear.
2. It operates at discrete times so that it can be handled by digital signal processing.
3. The system the filter acts on is stationary, i.e., no adaptation of the coefficients is necessary.

For the Wiener filter, the output $y(n)$, Eq. (2.4.41), is compared to the desired output $d(n)$. In order to find $d(n)$ one should precisely know the statistical properties of both signal and noise. The estimation error $e(n)$ is defined as the deviation of $y(n)$ from $d(n)$ [95]:

$$e(n) = d(n) - y(n). \tag{2.4.42}$$

Although the Wiener filter can be realized by an IIR filter, the following section is restricted to an FIR Wiener filter. Among different options to optimize the filter by minimizing the cost function J, the mean-square value of the estimation error $e(n)$ is selected. This is done because the mean-square value of $e(n)$ has a distinct minimum to uniquely point out the optimum statistical filter design [95]. The estimation error $e(n)$ taken at time n is a sample of a random variable. Therefore, the cost function J follows from the expectation value E of the squared absolute estimation error which is generally a complex quantity.

$$J = E\left[e(n)e^{*}(n)\right] = E\left[\left|e(n)\right|^{2}\right] \tag{2.4.43}$$

The gradient or nabla operator ∇ helps to find the minimum (ideally zero point) of the real-valued cost function J. As the filter coefficients h_r are also generally complex quantities, they can be rewritten in terms of their real part a_r and their imaginary part b_r.

$$h_r = a_r + \mathrm{j}b_r \tag{2.4.44}$$

The partial derivatives of J then lead to the complex gradient

$$\nabla J = \frac{\partial J}{\partial a_r} + \mathrm{j}\frac{\partial J}{\partial b_r}, \qquad r = 0,1,2,...R \tag{2.4.45}$$

If the filter is optimum, then ∇J must be equal to 0 for all $R + 1$ elements h_r. It should also be noted that J is a real function and independent of time. One can show that using Eq. (2.4.43) and Eq. (2.4.44) to compute the partial derivatives of Eq. (2.4.45) the following relation holds [95]:

$$\nabla J = -2E\left[u(n-r)e^{*}(n)\right] \tag{2.4.46}$$

For the optimum filter Eq. (2.4.46) is equal to zero.

$$E\left[u(n-k)e_o^{*}(n)\right] = 0 \tag{2.4.47}$$

The expression of Eq. (2.4.47) shows that $e(n)$ must be orthogonal to each corresponding input sample $u(n)$. With the estimation error of Eq. (2.4.41) and the filter output of Eq. (2.4.42) inserted in Eq. (2.4.47) the next relation is obtained [95]:

$$\sum_{i=0}^{R} h_i E\left[u(n-r)u^{*}(n-i)\right] = E\left[u(n-r)d^{*}(n)\right] \tag{2.4.48}$$

Two familiar terms can be identified on the left and on the right side of Eq. (2.4.48):

1. Autocorrelation term:

$$\vartheta_{uu}(i-r) = E\left[u(n-r)u^{*}(n-i)\right] \tag{2.4.49}$$

2. Cross-correlation term:

$$\vartheta_{ud}(-r) = E\left[u(n-r)d^*(n)\right] \tag{2.4.50}$$

These two relations are known as the Wiener-Hopf equations [95]. They define the optimum filter coefficients in terms of the autocorrelation function ϑ_{uu} of the filter input $u(n)$ and the cross-correlation ϑ_{ud} between filter input and desired response $d(n)$. Solving the Wiener-Hopf equations leads to an optimum set of filter coefficients. With these two relations, Eq. (2.4.48) can be rewritten:

$$\sum_{i=0}^{R} h_i \vartheta_{uu}(i-r) = \vartheta_{ud}(-r). \tag{2.4.51}$$

As an example for a Wiener filter, the so-called zero-forcing equalizer (Lucky et al, 1968) [95] is introduced. This filter is used to equalize, e.g., a dispersive channel where noise is not the dominant source of error. The equalizer combines the impulse responses of the channel with response $c(t)$ and of the equalizer with response $h(t)$ to form a single equivalent filter with the response $c(t) * h(t) = g(t)$. This combination is only valid if the system is fully linear. The coefficients h_r are selected such that a single impulse at the input produces a single impulse at the output leaving all other times with a value equal to zero, thus the term zero-forcing equalizer.

2.4.7.3 Adaptive Equalization

While the Wiener filter can only be applied to time invariant channels, adaptive equalization is often preferred. This is especially true if the properties of the signal and the channel characteristics are unknown or are subject to change. In this case a feedback loop adjusts the filter coefficients to optimize the filter performance. The most popular adaption algorithm is based on the least mean squares (LMS) [96]. One can distinguish between two processes, namely the filtering process and the adaptation of the coefficients h_r. Due to the feedback used for adapting h_r, the issue of stability arises. Therefore, during the filter design, it must be taken care of that the cost function J converges [95]. This method is similar to the Wiener filter as the same cost function J, see Eq. (2.4.43), is to be minimized. However, due to the adaptability of this filter, the final value of J_{LMS} will differ from J_{WIENER} obtained from the Wiener filter. This difference is referred to as excess mean-squared error. The ratio of excess mean-squared error and J_{WIENER} is called misadjustment. In contrast to the Wiener filter where the statistical properties of channel and signal have to be known, the LMS adapted filter estimates the gradient vector ∇J to adjust the filter coefficients, accordingly. One can show that the updated set of filter coefficients $h(n+1)$ are recursively obtained from the current coefficients $h(n)$ using the following relation [95]:

$$\begin{aligned} h(n+1) &= h(n) + \mu u(n)\left[d^*(n) - u^H(n)h(n)\right] \\ &= h(n) + \mu u(n)e^*(n) \end{aligned} \tag{2.4.52}$$

Here μ denotes the step-size parameter that controls the strength of the feedback and $u^H(n)$ is the transposed and complex conjugate of the input sample vector $u(n)$. The adaption process is started with an initial set of coefficients $h = h(0)$ that are corrected at each discrete time n. The capability of the LMS optimized filter to adapt to changes in the channel properties is ultimately limited by the highest possible value of μ that can be chosen while the system remains stable. In terms of processing requirements the LMS algorithm performs $2(R + 1) + 1$ complex multiplications and $2(R + 1)$ complex additions [95] and represents one of the simplest adaptive filter algorithms.

An LMS adapted equalizer [97] is supported in the Agilent Vector Signal Analyzer (VSA) software that is used throughout coherent communication experiments in this work. Two parameters have to be specified namely the filter order R or number of filter taps $R + 1$ and the step-size parameter μ here referred to as convergence parameter. Also the number of samples used per transmitted symbol influences the performance of the adaptation and equalization process. To demonstrate the power and substantial need for compensating the channel response, a 28 GBd QPSK signal has been coherently detected and results prior (a) and after (b) equalization are depicted in Fig. 2.43. A significant decrease of the constellation points' variances is to be seen. The probability of false symbol assignments is thus greatly reduced after equalization and so is the bit error probability.

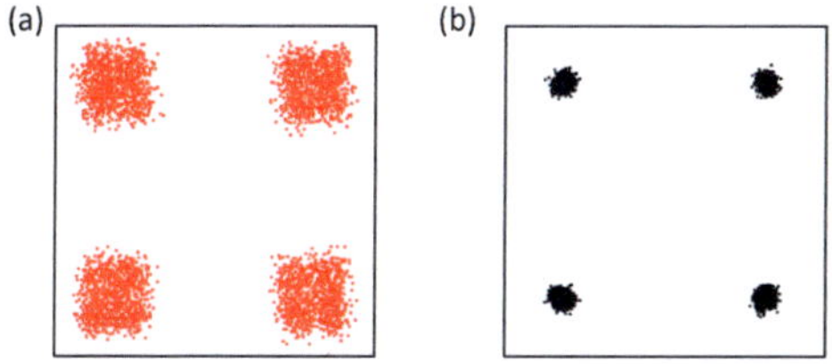

Fig. 2.43 Constellation diagrams of a coherently received 28 GBd QPSK signal. The variance of each constellation point prior to equalization (**a**) is significantly reduced after equalization (**b**). Therefore the equalized signal shows an enhanced robustness towards false symbol assignments.

As another advantage of adaptive equalization, details about the channel's frequency response can be drawn from the fully adapted and converged filter coefficients h_r. The absolute value of the channel response $G(f)$ (black) on a logarithmic scale along with the phase (red) is depicted in Fig. 2.44(a). As expected, the channel shows a low-pass character as higher frequencies experience higher attenuation. Mainly employed electronics such as amplifiers or data converters and the electrical modulation bandwidth of the optical I/Q-modulator determine the frequency roll-off. Fig. 2.44(b) depicts the real (blue) and imaginary (red) parts of the filter's impulse response h. The dots represent the actual filter coefficients h_r for an equalizer that operates with one sample per symbol which has proven best performance. In this example, the filter order R is equal to 44, i.e., there are $R + 1 = 45$ coefficients.

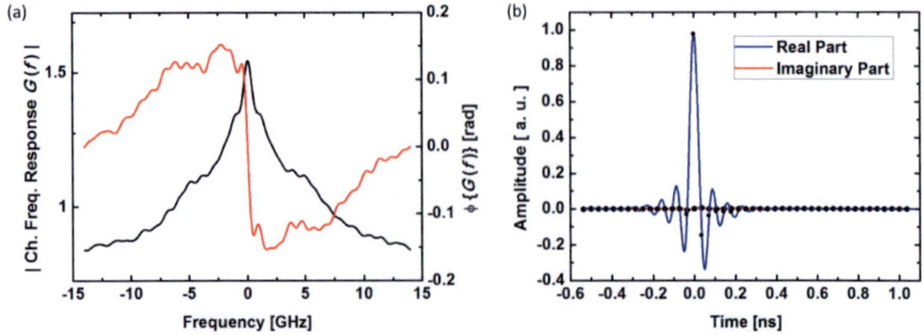

Fig. 2.44 (**a**) Determined channel response $G(f)$ of the QPSK transmitter and coherent receiver pair. The result is extracted from the converged filter coefficients h_r after the process of adaptation. From $|G(f)|$ (black, plotted on a logarithmic scale) it can be seen that the channel has low-pass character. The red curve represents the phase response $\Phi\{G(f)\}$ of the channel. (**b**) Equalizer impulse response h separated in real (blue) and imaginary (red) parts. Dots represent the actual filter coefficients h_r for an equalizer with one sample per symbol and a filter order $R = 44$.

3 Software-Defined Optical Transmitters

Software-defined optical Tx can generate a variety of different modulation formats and pulse-shapes at the push of a button. With this flexibility, the Tx can adapt to the changing situations in the network and is therefore able to optimally utilize the given resources such as bandwidth and maximum achievable signal-to-noise ratio (SNR). Furthermore, energy efficiency has become a major topic within the field of communications. With software-defined Tx and Rx, processing power can be flexibly saved in times of low network traffic. As another advantage of software-defined devices is the use of a common hardware platform which saves a reasonable amount of costs with respect to development and manufacturing.

The following section is structured as follows: First, the hardware used to build powerful digital signal processing circuits is introduced. Second, the system-on-chip design for the developed software-defined Tx is presented. Finally, the first experiments of a flexible real-time QAM Tx are discussed.

3.1 Field Programmable Gate Arrays

Digital signal processing at highest data rates, namely hundreds of gigabits, calls for sophisticated hardware. General purpose processors such as central processing units (CPU), do not offer the amount of resources needed and the energy efficiency is rather low. Standard microcontrollers are much too slow and even digital signal processors cannot handle datastreams of several hundreds of gigabits. Application specific integrated circuits (ASIC) are powerful and can be designed to meet all requirements. As a major downside, however, ASIC development is time-consuming and extremely costly. Furthermore, a fully designed ASIC cannot be easily adapted to meet requirements that may come up in the future. Especially in research, where changing circuit designs are the everyday situation by definition, the use of ASICs is not practical.

For prototyping of digital processors at highest data rates, field programmable gate arrays (FPGA) are the only practical solution. An FPGA offers a certain amount of logic cells arranged in logic blocks and memory blocks (usually referred to as block random access memory, BRAM) that can be programmed by the user a multitude of times. While programming, logic cells on the FPGA are connected through configurable "wires", depending on the desired functionality. Often high performance FPGAs also include special processing slices such as multipliers or hard-coded on-chip microcontrollers. FPGAs further offer high speed input-outputs (IO) and bidirectional data buses in order to interface the FPGA with external devices. In contrast to standard processors such as CPUs, where program code or software is executed during runtime line-by-line, the functionality of an

FPGA is realized in hardware (it is hard-wired). This strongly promotes the possibility of parallel processing in FPGA designs. So despite the fact, that the maximum frequency with which state-of-the-art FPGAs can be clocked is "only" in the range of 500 MHz, hundreds of gigabits can still be processed through parallelization.

The two major suppliers of FPGAs are Xilinx and Altera. The FPGAs used in this thesis belong to the Xilinx Virtex family.

3.1.1 History of Xilinx Virtex Familiy FPGAs

The very first family of Xilinx Virtex FPGAs has been introduced back in 1998 [98]. The biggest chip of that family offers a total number of 27,648 logic cells and up to 512 IOs [98]. Compared to state-of-the-art FPGAs as the ones found in the Virtex 6 and Virtex 7 families, the amount of resources back then seems very limited. And indeed the FPGA development over the last 15 years is quite remarkable. In order to give a better picture of this impressive development, Fig. 3.1 shows the maximum amount of logic cells provided by the biggest FPGA within each of the Virtex families together with the corresponding year at which the chip was first announced [98]–[104].

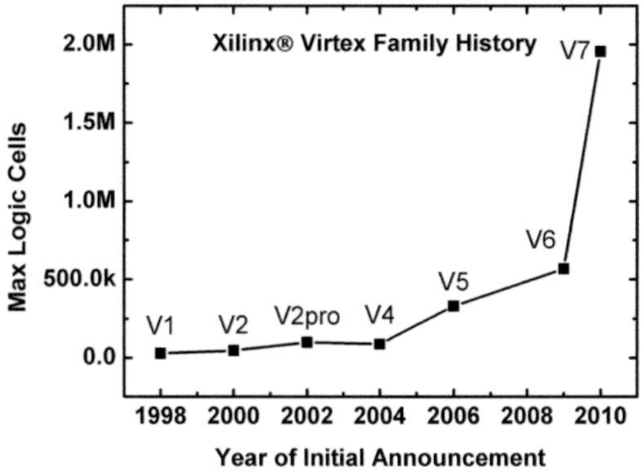

Fig. 3.1 Maximum amount of logic cells provided by the biggest Virtex FPGA of a certain family as a function of the year of initial announcement [98]–[104].

It can be seen that within the relatively short time of approximately one decade the amount of available logic on a Virtex FPGA has increased by several orders of magnitude. Especially the Virtex 7 family offers tremendous processing power which is expected to further grow for future Virtex families.

3.1.2 VHDL and Verilog

There are several possibilities to generate so-called bit-files that are used to program and therefore determine the functionality and behavior of FPGAs. Despite some high-level approaches where FPGA code is generated from Labview [105], Matlab [106], or even C/C++ [107], the most commonly used description languages are Verilog and VHDL (very high speed integrated circuit hardware description language). This is because high level descriptions ususally lead to large overhead and are typically not well suited for high speed designs. On the one hand, the Verilog language was introduced back in 1984, has a syntax that is closely related to the C programming language, and is mostly preferred by north American designers. On the other hand, VHDL was first introduced in the 1980s, has a syntax and structure closely related to the Ada programming language, and is mostly preferred by European designers. Both languages, Verilog and VHDL, can be allocated to the so-called dataflow languages [108] which are used to model the "ways of the data" within a digital circuit. This is in contrast to the family of procedural computing languages such as C/C++ that are used to describe sequential codes that are executed line-by-line.

There are several steps involved in order to generate a programmable bit-file from a Verilog/VHDL description. The first layer within the process is called the register transfer level (RTL). The RTL is a design abstraction that models synchronuous digital circuits in terms of the data flow. In this rather abstract level, simulations can be already performed using a so-called testbench that is usually also written in Verilog/VHDL. The process of synthesis (or logic synthesis) translates the RTL description into so-called netlists that contain the design implementation in terms of logic gates. These netlists are typically unspecific with respect to the target hardware. Possible timing constraints are not considered yet. As a last step the so-called place and rout process (PAR) uses the netlists together with further constraints with respect to timing and pin assignments to create a bit-file to program the target FPGA. Simulations of the code performed at this point reveal (next to the functionality) also the correct timing and thus maximum operation frequency of the created design in the target system.

3.2 System on Chip Design of the SW-Defined Tx

The system used as platform for the software-defined multi-format Tx was described with the VHDL language. The implemented functionalities can be coarsly divided into control tasks and data processing tasks. For the control tasks, also referred to as control flow, high versatility and flexibility at low-speeds is required whereas the processing or data flow must be performed at highest speeds. The system-on-chip design for the software-defined Tx is pictured in Fig. 3.2. A modular approach has been chosen in order to preserve high flexibility, easy maintenance and upgradeability.

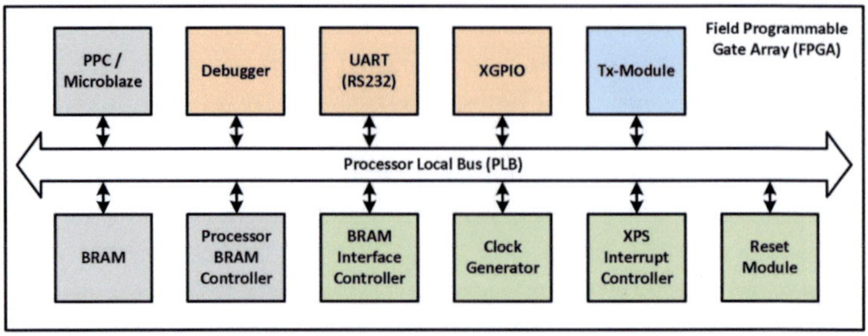

Fig. 3.2 Chip design on a system level for the SW-defined Tx.

All top-level modules are connected to a common so-called processor local bus (PLB). Most of the modules depicted in Fig. 3.2 contribute to the control flow with the main module being the power pc (PPC) for some Virtex FPGAs or the Microblaze for other Virtex FPGAs. This processor runs software (written in C-code) that is stored in the BRAM. In order to provide an interface to external devices, three modules (orange) have been implemented. The debugger is used to test the proper functionality of the written C-code. The universal asynchronuous Rx and Tx (UART) module is based on the RS232 standard and can be used to interact with the software-defined Tx from an external PC. Finally, the general purpose IO (XGPIO) is used to drive low-speed FPGA pins to communicate with peripheral hardware such as the DACs. Other modules (green) are used for example to provide an appropriate clock to the processor (Clock Generator) or to manage so-called interrupts issued e.g. by the UART. The data flow is handled by the Tx-Module where the actual high-speed DSP is performed.

The inside of the Tx-Module is depicted in Fig. 3.3. It can be seen that most of the modules can be controlled via registers (red lines) that are accessed by the processor. Furthermore, the modules are driven by a common clock (blue lines). First N pseudo random binary sequence (PRBS, left) generators produce bitstreams used to emulate real payload data. Due to the parallel nature of the FPGA design, each generator produces 32 bit per clock cycle. A multiplexer (MUX) then takes N arrays of 32 bit and forms an output sequence of $N \times 32$ bit. This binary sequence is fed to the Modulation Module where data is encoded according to the chosen modulation format (e.g., M-ary QAM) and pulse-shape (e.g. OFDM or Nyquist). The output of the Modulation Module contains a total of 128 samples with a resolution of 6 bit (128×6 bit = 768 bit) according to the DACs used in the target system. Therefore the number of samples being processed in parallel is 128 and the clock of the FPGA can be reduced by a factor of 128 as compared to the target sampling rate f_s of the DACs.

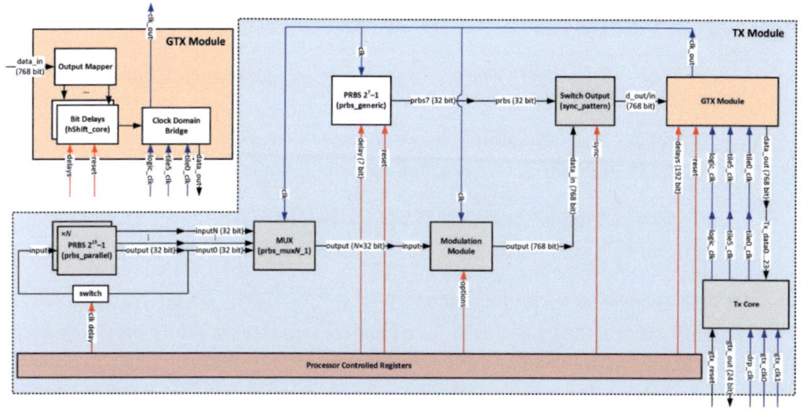

Fig. 3.3 Schematic design of the Tx_Module.

A second PRBS generator (prbs_generic) together with the Switch Output (sync-pattern) module are used to synchronize a number of 24 lines that feed the digital data to the DACs. With the Switch Output module, it can be selected whether the PRBS used for synchronization or the actual signal after the Modulation Module is to be transmitted. The next module is called GTX Module and provides several functionalities. A more detailed view on this module is given in the upper left of Fig. 3.3.

The Output Mapper rearranges the data prior to submission taking the hardware wiring between FPGA and DAC into account. A number of 24 Bit Delay modules have been implemented to correct for lengths mismatches of the 24 differential wires that connect FPGA and DAC. The values of these Bit Delay modules is correctly adjusted during the synchronization process. Finally a Clock Domain Bridge ensures that all processing blocks described so far can be operated with a single global clock and that all registers are stable when subsequent transmit the data to the DAC. The last module in the flow is the Tx Core. In order to keep the total number of wires between FPGA and DAC low, so-called multi-gigabit transceivers (MGT) of the FPGA are used. These transceivers support a maximum bitrate of 6.25 Gbit/s but have been operated at speeds of up to 7 Gbit/s [102]. For the Virtex 5 which has been used for the experiments in the course of this thesis, the MGTs are called GTX [102].

In the following an experiment with the described software-defined Tx is shown where various *M*-ary QAM signals have been generated at symbol rates of up to 28 GBd.

3.3 Real-Time Generation of 28 GBd 64QAM

Real-Time Software-Defined Multiformat Transmitter Generating 64QAM at 28 GBd

R. Schmogrow, D. Hillerkuss, M. Dreschmann, M. Huebner, M. Winter, J. Meyer, B. Nebendahl, C. Koos, J. Becker, W. Freude, and J. Leuthold

IEEE Photon. Technol. Lett. 22(21), pp. 1601–1603, (2010).

Today's high performance communication systems rely heavily on optical transmission links. High-speed electronics is crucial to exploit the large bandwidth of optical systems. So far, optical backbone networks were operated mostly with pulse amplitude modulation (PAM) and phase-shift keying (PSK) modulation formats such as differential PSK (DPSK) and differential quadrature PSK (DQPSK) [109]. However, future optical networks will operate with multi-level coded signals such as *M*-QAM [110]. Advanced modulation formats promise enhanced spectral efficiency at the cost of more complex transmitters and receivers. There are several ways to implement quadrature amplitude modulation (QAM) transmitters, such as discrete electrical digital-to-analog converters (DAC) [111], optical multi-modulator schemes [112], all optical DACs [113], and integrated electrical DACs in the form of arbitrary waveform generators (AWG). Although the AWG is the most versatile solution, its capability is limited due to finite memory size, and due to the lack of real-time processing. Therefore a more powerful solution has to be found. Field programmable gate arrays (FPGA) are able to handle the required amount of data, and yet offer the flexibility to change their functionality through software. Combined with state-of-the-art high speed DACs, a software defined transmitter can be implemented. With such a scheme, real-time Nyquist sampling for pre-compensating dispersion at 10.7 Gbit/s was demonstrated [65]. Recently, we have introduced a highly flexible and synchronous transmitter for modulation formats as complex as 16QAM [114], providing online data generation and real time digital signal processing at the same time.

3.3.1 Experimental Setup

The experimental setup of the software-defined transmitter comprises several electrical and optical components as illustrated in Fig. 3.4. An external cavity laser source provides the optical carrier to be modulated in nested $LiNbO_3$ I/Q-Mach-Zehnder modulators (MZM, electrical bandwidth ≈ 28 GHz, π-phase shift voltage $V_\pi \approx 2$ V). The electrical signal is generated by two MICRAM high-speed DACs, the outputs of which are amplified for driving the MZM. Both DACs are supplied with an electrical clock with a maximum frequency of 28 GHz. A variable electrical phase shifter aligns the two DAC outputs in phase with respect to each other. Xilinx Virtex5 field pro-grammable gate arrays (FPGA) drive the DACs, each of which providing 24 over-clocked feeding lines operating

at up to 7 Gbit/s each. The over-clocking did not cause stability issues. The feeding lines are 4:1 multiplexed by the DAC, resulting in an overall symbol rate of up to 28 GBd with a resolution depth of 6 bit. The electrical clock for the FPGA is generated by frequency dividers located on the DAC board. Real-time com-putation is performed by the FPGA devices incorporating both, signal generation, or external data accommodation and signal processing. To emulate dual polarization (DP) signals, the modulated signal is split, and a delay of 5.3 ns is applied to one of the paths for de-correlation. A variable optical attenuator equalizes the optical power in the two paths. The orthogonally polarized signals are then combined.
To judge the quality of the transmitter, an Agilent N4391A Optical Modulation Analyzer (OMA) receives, post-processes, and analyzes the constellations. Further, an Agilent Digital Communications Analyzer (DCA) measures eye diagrams and detects intensities. In order to receive sufficient optical power in the two units, the signal was amplified by an erbium doped fiber amplifier (EDFA), and then optically band pass filtered to suppress the amplified spontaneous emissions generated by the EDFA. The signal is split and fed to the OMA and the DCA. Intensities were measured for each polarization sepa-rately. The decoded data served for bit error ratio (BER) and error vector magnitude (EVM) measurements. We recorded constellation diagrams for 8 different modulation formats.

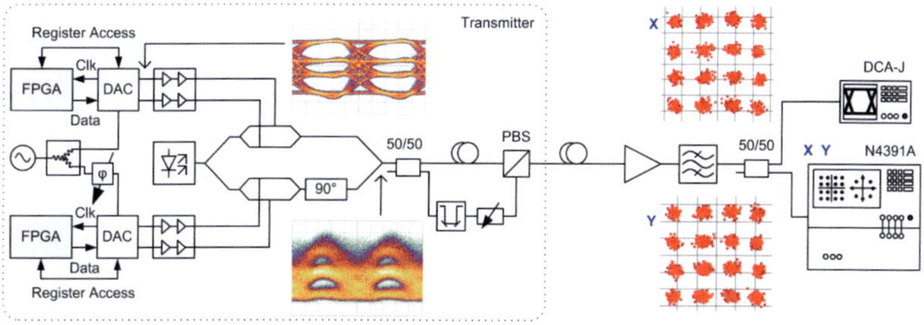

Fig. 3.4 Experimental setup and transmitter implementation. A semiconductor laser source is I/Q modulated by nested Mach-Zehnder modulators. Each arm is driven by the amplified output of two DAC that are both clocked at 28 GHz. Two FPGA are used to drive the DAC and perform real-time signal processing. DAC register access is needed for synchronization. A 2^{15}-1 PRBS is generated on both FPGA resulting in a four level electrical output, which was pre-distorted to compensate for the nonlinear transfer function of the modulator. The output of the I/Q modulator is polarization multiplexed, amplified, and then split. A DCA with a 50 GHz optical input was used to measure intensity, and the N4391A OMA decoded the DP-16QAM signal resulting in **X** and **Y**.

With an Agilent OMA we measured the EVM and BER of various formats and symbol rates. For convenience, we chose a 2^{15}-1 PRBS as data source. BER calculated according to [115] are denoted with a prefix '~' as opposed to measured BER. Single Polarization 28 GBd experiments were performed first. For BPSK and QPSK EVMs of 5.7 % (BPSK) respectively 8.1 % (QPSK) were found, resulting in a BER below measurement limits.

16QAM exhibited errors below a BER of ~ 1×10^{-7} with an EVM of 8.6 %. 8PSK (EVM = 12.5 %), 32QAM (EVM = 9.1 % / BER ~ 3.5×10^{-4}), and 64QAM (EVM = 10.1 % / BER ~ 9.0×10^{-3}) were also measured at 28 GBd. For DP-16QAM we found a total BER of 5.3×10^{-5}. We determined an EVM of 10.2 % for one, and 10.7 % for the other polarization. The remaining formats were recorded at 20 GBd, namely 4PAM and 6PAM.

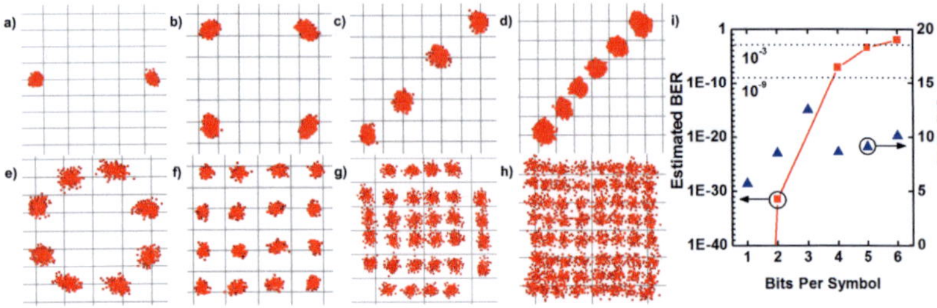

Fig. 3.5 Received and decoded modulation formats with corresponding constellation diagrams. (**a**) BPSK at 28 GBd with an EVM of 5.7 % exhibited no errors, (**b**) QPSK at 28 GBd showed no errors with an EVM of 8.5 %, (**c**) 4PAM at 20 GBd, (**d**) 6PAM at 20 GBd, (**e**) 8PSK at 28 GBd with an EVM of 12.5 %, (**f**) 16QAM at 28 GBd with an EVM of 8.6 % and no errors, (**g**) 32QAM at 28 GBd with a resulting EVM of 9.1 %, (**h**) 64QAM at 28 GBd with an EVM of 10.1 %, (**i**) Calculated BER values and measured EVMs for QPSK, 16, 32 and 64QAM formats. FEC limits are indicated for a BER of 10^{-9} and 10^{-3}.

The software-defined transmitter can generate 8 different modulation formats at symbol rates up to 28 GBd. A total of 48 feeding lines between FPGAs and DACs were synchronized resulting in 336 Gbit/s. Dynamic format switching in only 5 ns was performed through the processor-user interface by software adjustable LUTs. No hardware reconfiguration was needed for this purpose. Manual re-adjustment is only required when changing the symbol rates or when switching to on-off modulation formats. Advanced modulation schemes such as DP-16QAM at 224 Gbit/s were successfully tested resulting in bit error ratios well below the FEC limit of BER = 2×10^{-3}. A variety of different modulation schemes (Fig. 3.5) including 32QAM and 64QAM were experimentally demonstrated. When using the described transmitter at 64QAM in a dual polarization setup, data-rates of up to 336 Gbit/s on a single optical carrier can be achieved.

4 Real-Time Digital Pulse-Shaping

Pulse-shaping techniques such as OFDM or Nyquist signaling have proven the potential to significantly enhance the performance of optical transmission links. At the same time, real-time digital processing has become of major interest within the field of optical communications. In this section, first OFDM and Nyquist pulse-shaping are compared from a theoretical point-of-view. Second, real-time Nyquist Tx are presented that are able to generate Nyquist sinc-pulses at data rates > 100 Gbit/s. Third, real-time OFDM transmitters in the 100 Gbit/s regime are discussed. Finally, FPGA implementations of FPGA based Nyquist and OFDM Tx are analyzed in terms of resources and their performance is evaluated for single carrier and multi-carrier optical transmission.

4.1 OFDM and Nyquist Pulse-Shaping: A Comparison

Real-time Nyquist pulse generation beyond 100 Gbit/s and its relation to OFDM

R. Schmogrow, M. Winter, M. Meyer, A. Ludwig, D. Hillerkuss, B. Nebendahl, S. Ben-Ezra, J. Meyer, M. Dreschmann, M. Huebner, J. Becker, C. Koos, W. Freude, and J. Leuthold

Opt. Express 20(1), pp. 317–337 (2012). [J22]

Sinc-shaped Nyquist pulses spread into adjacent time slots, but their rectangularly shaped spectra require only the minimum Nyquist channel bandwidth. They are well known from communication theory but are relatively new in optical communications. The Nyquist modulation format is very similar to optical orthogonal frequency division multiplexing (OFDM), where sinc-shaped sub-spectra extend into adjacent frequency slots, and symbols in time are rectangularly shaped. In the course of this paper all Nyquist pulses are sinc-shaped.

Here, we first discuss the close relation of Nyquist pulse modulation with OFDM [23][116][117]. Both Nyquist pulse shaping and OFDM are described with a similar formalism. This way it will become clear that Nyquist modulation is nothing but an orthogonal time division multiplexing technique, much the same as OFDM is an orthogonal frequency multiplexing technique. Furthermore, we compare the two multiplexing methods with respect to their characteristics like spectral efficiency (SE) and peak-to-average power ratio (PAPR). We then demonstrate real-time Nyquist pulse generation for signals beyond 100 Gbit/s. This has become possible even with the limited speed of state-of-the art electronics [40]. In more detail, we generate quadrature phase shift keying (QPSK) at 56 Gbit/s and quadrature amplitude modulation with 16 states (16QAM) at 112 Gbit/s in combination with polarization division multiplexing (PDM). This results in an overall spectral efficiency of 7.5 bit/s/Hz for PDM-16QAM. Finally the reception of Nyquist shaped pulses is discussed comparing it with the reception of standard non-return-to-zero (NRZ) QAM signals.

4.1.1 Advanced filtering in optical WDM networks

Modern optical networks rely on multi-wavelength and multi-carrier transmission systems in order to fully exploit the bandwidth offered by optical fibers. The ultimate target is to maximize the spectral efficiency, i.e., the amount of transmitted data within a given bandwidth [118]. In general, the maximum capacity of a channel is only limited by Shannon's law. For optical communications non-linear distortions limit the ultimate channel capacity at high launch powers. Thus increasing the signal to noise ratio (SNR) by increasing the signal power is only possible within certain limits [119][120]. For high capacity networks, coming close to this so-called non-linear Shannon limit is of special interest.

For conventional M-ary QAM signals, the spectral occupancy does not alter significantly when changing the number of bits b transmitted per symbol. Thus increasing the number of constellation points $M = 2^b$ leads directly to an increase in spectral efficiency. However, transmitting an additional bit per symbol implies doubling the number of constellation points, so that for a constant average power the required signal-to-noise ratio (SNR) increases significantly. This is also true if the spectral efficiency is increased by polarization division multiplexing (PDM) or polarization switching [78].

Bandwidth can be saved, however, when applying advanced filtering. From signal theory we know that the minimum bandwidth needed to fully encode a bandwidth-limited signal is the Nyquist bandwidth F_s [61]. If the optimization of spectral efficiency is the ultimate target, all frequency components outside the Nyquist band must be removed by filters. As a consequence, the time domain signal changes from pulses that are clearly separated in time (e.g., non-return-to-zero format, NRZ) to pulses that overlap their neighbors.

As an example, Fig. 4.1(a) displays the spectrum of an M-ary QAM NRZ signal for three different WDM channels centered at optical frequencies f_0, f_1, and f_2. The spectra are significantly wider than F_s, but can be reduced to the Nyquist bandwidth without loosing any signal information. However, appropriate filtering is required to achieve the best possible transmission quality. The sinc-shaped spectrum of an NRZ signal should be filtered such that the resulting spectrum is of rectangular shape under the assumption that the frequency response of the channel is flat in the region of interest. Therefore the side lobes must be removed, and the spectrum within F_s must be flattened. If there are slopes in the channel's frequency response, or the noise accumulated in the system is not constant over frequency, a pre- and de-emphasis filtering scheme should be applied. In a properly filtered WDM spectrum comprising the same three carrier wavelengths as in Fig. 4.1(a), the channels can now be placed next to each other located on a frequency grid the minimum spacing of which is dictated by the symbol rate F_s (Fig. 4.1(b), Nyquist-WDM [24][31]).

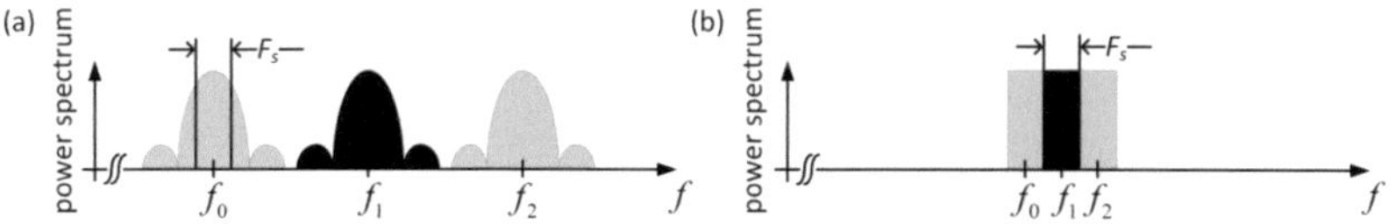

Fig. 4.1 Reducing the required channel spacing in WDM systems by removing spectral portions outside the Nyquist bandwidth F_s through filtering. Minimum channel spacing for crosstalk free systems is fixed by F_s. (**a**) Unfiltered M-ary QAM channels with theoretically infinitely wide spectrum. (**b**) Nyquist channel center frequencies can be spaced apart by the Nyquist bandwidth F_s. Basically, no guard bands are required.

In general, the previously described filters can be implemented optically, electrically, or digitally. Possible implementations are shown for a software-defined transmitter, Fig. 4.23. Optical filters with a transfer function $S_{21}(f)$ as in Fig. 4.23(a) could be used. The difficulty is to build optical filters with frequency responses that drop significantly inside just a few MHz. Optical filters based on liquid crystals may offer an opportunity to perform such filtering [121]. Nevertheless, these filters are quite elaborate and show some penalties due to the limited slopes in their frequency response. Electrical filters as shown in Fig. 4.23(b) are another option. They can provide very steep slopes. A complex transmitter, however, requires two filters with a specific frequency transfer function depicted in Fig. 4.23(b). As before, these analog electrical filters are not easily available. Conversely, designing digital filters to be included in the digital signal processing (DSP) part of the transmitter [19] seems to be a suitable option to solve the problem. State-of-the-art software-defined optical transmitters [73] utilize DSP functionality which only has to be extended. Naturally, digital filtering calls for additional analog anti-aliasing filters to remove image spectra. These filters can be standard low-pass filters as any negative influence can be pre-compensated by digital filtering. Furthermore, only DSP offers the flexibility to vary filter coefficients during runtime and therefore the capability to adapt to changes in the channel response. Additionally, changing the symbol rate F_s of the digital filter based transmitter is achieved without changing the hardware. Analog filters are generally fixed with respect to their frequency responses and cannot easily be altered.

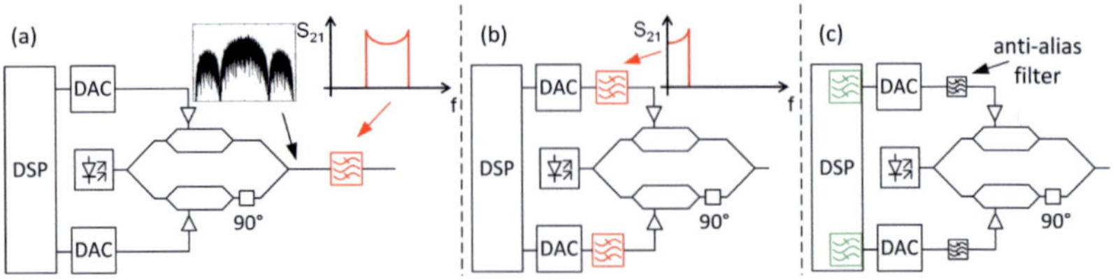

Fig. 4.2 Pre-filtering of optical *M*-QAM signals for generating sinc-shaped Nyquist pulses. (**a**) An optical filter can be applied to carve an ideal rectangle out of an NRZ spectrum. Such an optical filter requires a frequency response $S_{21}(f)$ that would be difficult – if not impossible – to realize. (**b**) Alternatively, two analog electrical filters can be used to form an appropriate output signal with an ideal rectangularly shaped spectrum. These filters would also need a non-standard transfer function $S_{21}(f)$. (**c**) By digital filtering as a part of the digital signal processing (DSP) block in the transmitter one can efficiently remove redundant parts of the spectrum. Resulting signals show almost ideal rectangular spectra, and additional off-the-shelf analog electrical or optical filters can easily remove spurious spectra.

4.1.2 Nyquist pulse modulation and OFDM: A comparison

Nyquist pulse modulation can be derived from the well-known optical orthogonal frequency division multiplexing (OFDM) technique [23]. This is done by simply interchanging time and frequency domain when describing the signal.

In general, an OFDM signal $x(t)$ is an *infinite* sequence of *temporal* symbols $x^{(i)}(t)$ superscripted with i. Each temporal symbol consists of a superposition of N *temporal sinusoidals* with equidistant carrier frequencies f_k inside a temporally rectangular window defining the temporal symbol length T_s. Frequency spacing $F_s = f_{k+1} - f_k = 1 / T_s$ and temporal symbol length T_s are interrelated to establish orthogonality, Eq. (6.2.2) in the Appendix. To simplify the discussion, we let aside a possible cyclic prefix that would reduce the symbol rate below F_s, and would therefore increase the temporal symbol spacing to a value larger than T_s. The OFDM carriers are encoded with complex coefficients c_{ik}. We find for the OFDM signal

$$x(t) = \sum_{i=-\infty}^{+\infty} x^{(i)}(t), \quad x^{(i)}(t) = \sum_{k=0}^{N-1} c_{ik} x_k(t - iT_s),$$
$$x_k(t) = \mathrm{rect}\left(\frac{t}{T_s}\right) e^{j2\pi f_k t}, \quad |f_{k+1} - f_k| = F_s = \frac{1}{T_s}. \tag{2.5.1}$$

The rectangular function rect(z) is 1 for $|z| < 1/2$ and zero otherwise, see Eq. (6.2.1) in the Appendix. By Fourier transforming Eq. (2.5.1) we obtain the frequency domain representation of the i-th temporal OFDM symbol, i.e., a set of N spectral sinc-functions centered at frequencies f_k,

$$X(f)=\sum_{i=-\infty}^{+\infty}X^{(i)}(f),\quad X^{(i)}(f)=\sum_{k=0}^{N-1}c_{ik}X_k(f)\mathrm{e}^{-\mathrm{j}2\pi f i T_s},$$
$$X_k(f)=T_s\,\mathrm{sinc}\left(\frac{f-f_k}{F_s}\right). \tag{2.5.2}$$

In contrast to OFDM, the *spectrum* $Y(f)$ of a Nyquist signal is a *finite* sequence of N *spectral* symbols superscripted with i. Each spectral symbol consists of a superposition of *infinitely* many *spectral sinusoidals* with equidistant Nyquist pulse position times t_k ("carrier" positions) inside a spectrally rectangular window defining the spectral symbol length F_s. Temporal spacing $T_s = t_{k+1} - t_k = 1 / F_s$ and spectral symbol length F_s are interrelated to establish orthogonality, Eq. (6.2.3) in the Appendix. The Nyquist "carriers" are again encoded with complex coefficients c_{ik}. In analogy to Eq. (2.5.1) we find

$$Y(f)=\sum_{i=0}^{N-1}Y^{(i)}(f),\quad Y^{(i)}(f)=\sum_{k=-\infty}^{+\infty}c_{ik}Y_k(f-iF_s),$$
$$Y_k(f)=T_s\,\mathrm{rect}\left(\frac{f}{F_s}\right)\mathrm{e}^{-\mathrm{j}2\pi f t_k},\quad |t_{k+1}-t_k|=T_s=\frac{1}{F_s}. \tag{2.5.3}$$

By Fourier transforming Eq. (2.5.3) we obtain the time domain representation of the ith spectral Nyquist symbol, i.e., a set of infinitely many temporal sinc-functions centered at times t_k,

$$y(t)=\sum_{i=0}^{N-1}y^{(i)}(t),\quad y^{(i)}(t)=\sum_{k=-\infty}^{+\infty}c_{ik}y_k(t)\mathrm{e}^{+\mathrm{j}2\pi i F_s t},$$
$$y_k(t)=\mathrm{sinc}\left(\frac{t-t_k}{T_s}\right). \tag{2.5.4}$$

The relations Eq. (2.5.1) – (2.5.4) are visualized in Fig. 4.3. The left column, Fig. 4.3(a) and (c), describes the time-frequency correspondence for OFDM, while the right column, Fig. 4.3(b) and (d), relates to Nyquist pulses. The upper rows of each section in Fig. 4.3 show the time dependency of the signals, while the lower rows refer to the corresponding spectra.

In Fig. 4.3(a), three temporally sinusoidal subcarriers modulated with $c_{i1} = c_{i2} = c_{i3} = 1$ form a specific OFDM symbol with width T_s and positioned at $t = 0$. The OFDM spectrum is a superposition of three spectral sinc-functions located at frequencies f_{k-1}, f_k, and f_{k+1}, which are separated by F_s. In Fig. 4.3(b), the superposition of three temporal sinc-functions is seen which are located at times t_{k-1}, t_k, and t_{k+1} and separated by T_s. These Nyquist pulses are modulated with $c_{i1}= c_{i2}= c_{i3} = 1$ and form a specific spectral Nyquist symbol with width F_s and position at $f= 0$. It consists of three spectrally sinusoidal Nyquist "subcarriers". The graphs in Fig. 4.3(a) and (b) represent Eq. (2.5.1) – (2.5.4) for $i = 0$, i.e., for an OFDM and a Nyquist symbol positioned at $t = 0$ and $f= 0$, respectively.

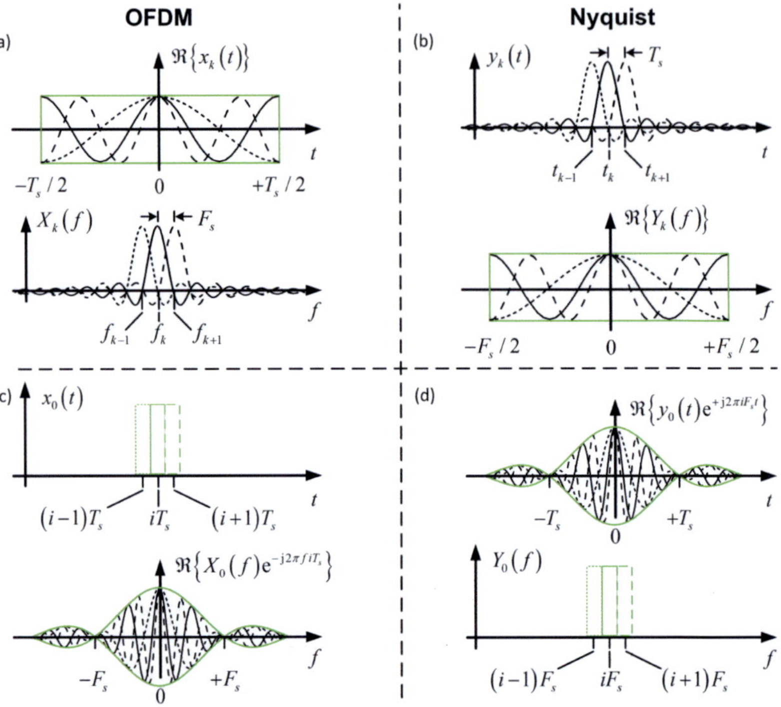

Fig. 4.3 Comparison between OFDM (left column) and Nyquist sinc-pulses (right column) in time and frequency domain. (**a**) The upper graph shows the real parts of three on-off keyed sinusoidal subcarriers $x_k(t)$, the sum of which represents one specific time-domain OFDM symbol centered at $t = 0$. The lower graph shows the corresponding spectra $X_k(f)$ of the subcarriers centered at frequencies f_{k-1}, f_k, and f_{k+1}. (**b**) The upper graph shows three Nyquist pulses $y_k(t)$ ("temporal subcarriers") centered at times t_{k-1}, t_k, and t_{k+1}. The lower graph shows the corresponding real parts of the spectra $Y_k(f)$ with three spectral sinusoidals, the sum of which represents one specific frequency-domain Nyquist symbol centered at $f = 0$. (**c**) The upper graph shows the envelopes (green rectangles as in Fig. 4.3(a) of three temporal OFDM symbols located at times $(i-1)T_s$, iT_s, and $(i+1)T_s$. For simplicity, each temporal OFDM symbol is composed of the same single zero-frequency subcarrier $f_0 = 0$ for $k = 0$, The lower graph shows the corresponding real parts of the spectra within a sinc-shaped envelope (green). The three spectral sinusoidals within correspond to three temporal positions of the temporal OFDM symbol. (**d**) The upper graph shows the real parts of three Nyquist pulses within a sinc-shaped envelope (green). The three temporal sinusoidals within correspond to three spectral positions of the spectral Nyquist symbols. The lower graph shows the envelopes (green rectangles as in Fig. 4.3(b) of the three spectral Nyquist symbols located at frequencies $(i-1)F_s$, iF_s, and $(i+1)F_s$. For simplicity, each spectral Nyquist symbol is composed of the same single zero-time Nyquist "subcarrier" $t_0 = 0$ for $k = 0$.

If we set $k = 0$, then each of the three OFDM or Nyquist symbols shown here consists of only one temporal zero-frequency ($f_0 = 0$) or spectral zero-time ($t_0 = 0$) "sinusoidal", respectively. For OFDM, the three temporal symbols are positioned at times $(i-1)T_s$, iT_s and

$(i+1)T_s$, Fig. 4.3(c). The resulting spectrum is located within a sinc-shaped envelope having its first zeros at $-F_s$ and $+F_s$. Due to the different positions of the temporal symbols we see three spectral sinusoidals within the (green) spectral envelope. For Nyquist pulses, the three temporal sinusoidals inside the (green) sinc-shaped pulse envelope with zeros at $-T_s$ and $+T_s$ correspond to three spectral symbols positioned at frequencies $(i-1)F_s$, iF_s, $(i+1)F_s$, Fig. 4.3 (d).

A schematic of OFDM signal and Nyquist pulse generation is given in Fig. 4.4. The left column, Fig. 4.4(a) and (c), refers to OFDM, whereas the right column, Fig. 4.4(b) and (d), describes Nyquist pulse generation. For a better understanding we set one of the summation variables k or i of Eq. (2.5.1) – (2.5.4) to zero while varying the other one, and we present the signal generation in both frequency and time domain.

For OFDM signal generation in the frequency domain, Fig. 4.4(a), a real sinc-shaped spectrum $X_{k=0}(f)$ centered at $f = 0$ is shifted by a finite number of equidistant frequency steps kF_s, $k = 0\ldots N-1$. The resulting sub-spectra are modulated by complex coefficients c_{ik}. The total OFDM spectrum $X^{(0)}(f)$ for $i = 0$ is formed by superimposing all N subcarrier spectra (Σ stands for summation), resulting in an OFDM symbol located at $t = 0$ only.

For Nyquist pulse generation in the time domain, Fig. 4.4(b), a real sinc-shaped impulse $y_{k=0}(t)$ centered at $t = 0$ is shifted by an infinite number of equidistant time steps kT_s, $k = -\infty\ldots+\infty$. The impulses are modulated by complex coefficients c_{ik}. The total Nyquist pulse $y^{(0)}(t)$ for $i = 0$ is formed by superimposing all "subcarrier" pulses, resulting in a Nyquist pulse sequence at one carrier "frequency" $f = 0$ only.

For OFDM pulse generation in the time domain, Fig. 4.4(c), a real rect-shaped pulse $x_{k=0}(t)$ comprising only one carrier "frequency" $f = 0$ is shifted by an infinite number of equidistant time steps iT_s, $i = -\infty\ldots+\infty$. These sub-pulses are modulated by complex coefficients c_{ik}. The total OFDM time signal $x(t)$ for $f = 0$ is formed by superimposing infinitely many temporal sub-pulses.

For Nyquist signal generation in the frequency domain, Fig. 4.4(d), a real rect-shaped spectrum $Y_{k=0}(f)$ comprising only one Nyquist pulse ("carrier") at $t = 0$ is shifted by a finite number of equidistant frequency steps iF_s, $i = 0\ldots N-1$. The resulting sub-spectra are modulated by complex coefficients c_{ik}. The total Nyquist symbol $Y(f)$ at $t = 0$ is formed by superimposing all N sub-spectra.

OFDM and Nyquist *receivers* can be built similar to the transmitter scheme depicted in Fig. 4.4. To this end, the received signal would enter from the right, the symbol Σ would represent a splitter, and local oscillators with complex conjugate time dependency (OFDM signal) or complex conjugate Nyquist pulses (Nyquist signal) mix with the incoming signals to recover the modulation coefficients c_{ik} having integrated over the symbol period T_s (for OFDM signal) or over all times (for Nyquist signals). Forming the complex conjugate means reverting the signs of frequency steps F_s and time steps T_s, respectively.

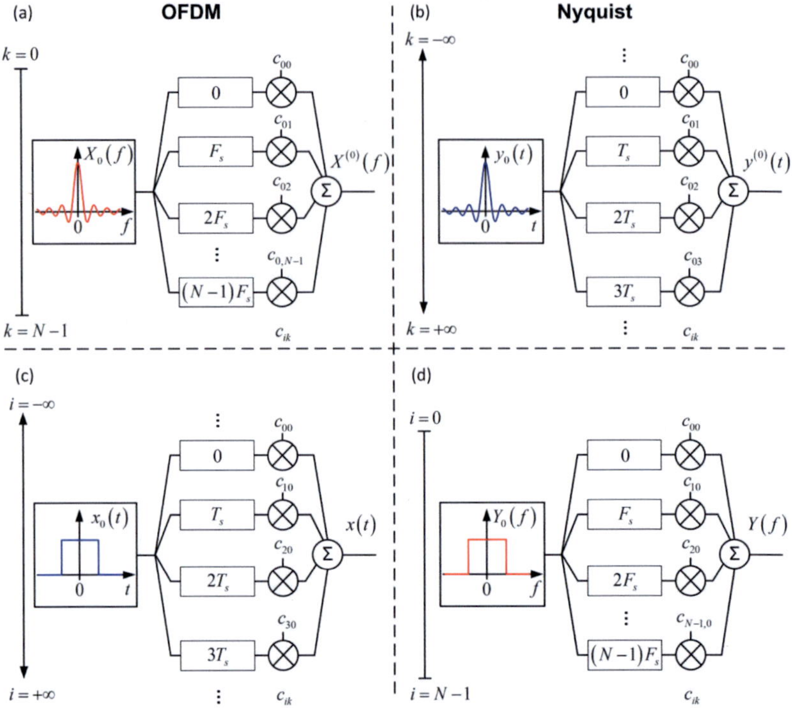

Fig. 4.4 Schematic of OFDM signal and Nyquist pulse generation, both in frequency domain and time domain. For a better understanding we keep either k or i of Eq. (2.5.1) – (2.5.4) constant at zero while varying the other quantity. (**a**) OFDM spectrum, OFDM symbol at $t_0 = 0$ ($i = 0$) only: N sinc-shaped sub-spectra for a finite number of $k = 0 \ldots N-1$ keyed subcarriers are frequency shifted by kF_s and modulated with complex coefficients c_{ik}. The superposition of all modulated sub-spectra results in the total OFDM spectrum $X(f) = X^{(0)}(f)$. (**b**) Nyquist pulse, Nyquist symbol with $f_0 = 0$ ($i = 0$) only: Infinitely many sinc-shaped pulses (Nyquist "subcarriers", $k = -\infty \ldots +\infty$) time shifted from $t = 0$ by increments kT_s and modulated with complex coefficients c_{ik}. The superposition of all modulated pulses results in the total Nyquist time signal $y(t) = y^{(0)}(t)$. (**c**) OFDM symbol, OFDM spectrum with $f_0 = 0$ ($k = 0$) only: Infinitely many rect-shaped temporal pulses ($i = -\infty \ldots +\infty$) time shifted from $t = 0$ by increments iT_s and modulated with complex coefficients c_{ik}. The superposition of all modulated pulses results in the total OFDM time signal $x(t)$. (**d**) Nyquist symbol, Nyquist pulse at $t_0 = 0$ ($k = 0$) only: N rect-shaped sub-spectra for a finite number of $i = 0 \ldots N-1$ keyed spectral symbols are frequency shifted by iF_s and modulated with coefficients c_{ik}. The superposition of all modulated spectra results in the total Nyquist spectrum $Y(f)$.

An in-depth mathematical comparison between OFDM and Nyquist pulse shaping is given in the Appendix. Due to the close relation to OFDM, Nyquist pulse generation could be also referred to as an orthogonal *time* division multiplexing (OTDM) technique.

4.1.3 Oversampled Nyquist pulses with finite-length

An elementary Nyquist shaped impulse with minimum spectral width is a sinc-function infinitely extended in time. Real Nyquist pulses, however, need to be approximated by a finite-length representation. For practical reasons finite impulse response (FIR) filters are used to build the pulse shapes [122][123]. In addition, for separating the baseband spectrum from its periodic repetitions using realizable filters, oversampling by a factor q (typically $q = 1.2, 2, \ldots$) is needed. In this paper we have chosen $q = 2$. This way we will subsequently save FPGA resources since sampling points of adjacent symbols fall onto the same time slot. However, smaller oversampling factors such as $q = 1.2$ suffice if adequate anti-aliasing filters are available. This would allow us to reduce the required processing speed and DAC sampling rate but comes at the cost of an increased processing complexity.

A suitable FIR filter of order R can be constructed by a sequence of R delay elements T_s / q with $T_s = 1 / F_s$, and $R + 1$ taps in-between. The tapped signals are weighed by a number of R so-called filter coefficients h_r and summed up to form the filter output, Fig. 4.5. A "one-tap" filter with order $R = 0$ reproduces the filter input.

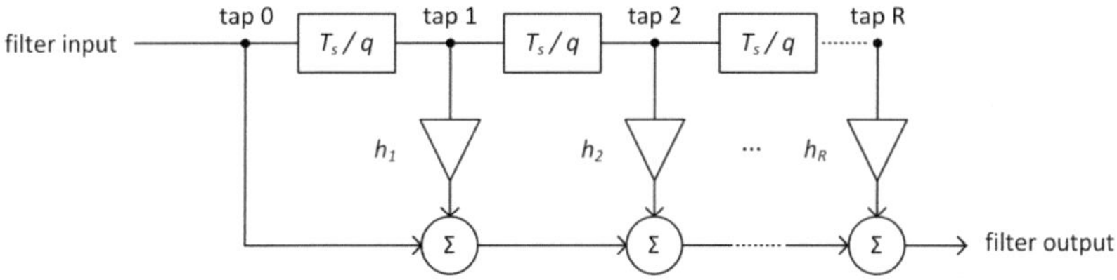

Fig. 4.5 Finite impulse response filter (FIR, direct form I) of order R. A series of R delay elements T_s / q are located in-between the $R + 1$ taps. Tapped signals are weighed by R filter coefficients h_r and summed to form the filter output.

Signal generation with various FIR filter orders R is shown in Fig. 4.6. The left column shows the impulse response of each filter. The effective windowing is indicated by a green rectangle. The linearly scaled corresponding transfer functions are seen in the middle column. The right column displays these same transfer functions on a logarithmic scale. The spectra of the single pulses (white lines) are plotted together with simulated data (colored). A two-fold oversampling $q = 2$ is used in this context.

The simulation was performed as follows: A pseudo random binary sequence (PRBS) with a length of $2^{15} - 1$ serves as origin for simulated complex data. As a reference, these complex data c_{ik} modulate NRZ pulses, one of which is displayed in Fig. 4.6(a), left column. The linearly scaled sinc-shaped power spectrum of this elementary impulse is seen in Fig. 4.6(a), middle column. The logarithm of the same power spectrum is shown as a white line in Fig. 4.6(a), right column, together with the ensemble-averaged power spectrum for the simulated data. For all power spectra a possibly existent discrete carrier line is omitted.

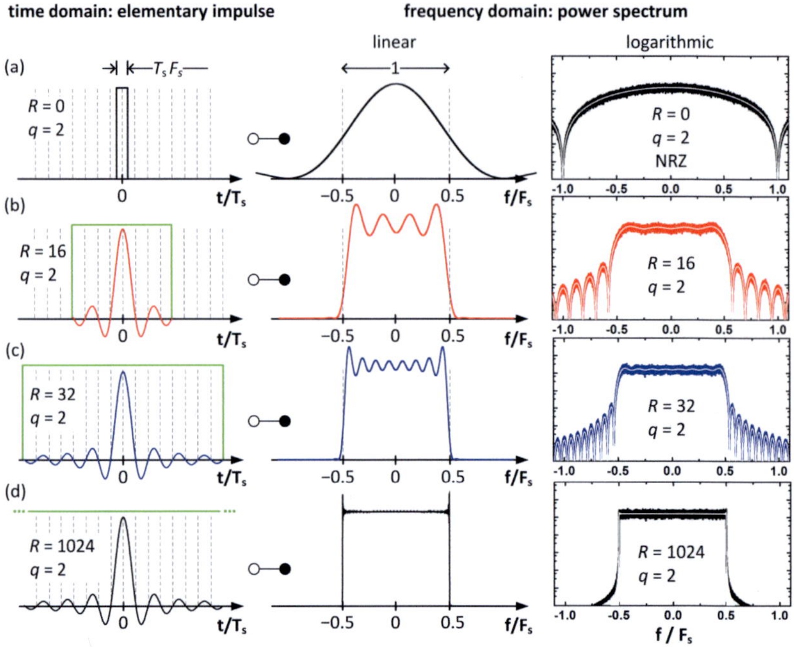

Fig. 4.6 Impulse responses and transfer functions of FIR filters with various orders R. Left column: Impulse responses. The durations of the impulse response is marked with a rectangular window (green). Middle column: Power spectra on a linear scale. Right column: Power spectra on a logarithmic scale. Colored noisy curves represent the average power spectra for a pulse train with a repetition period $T_s = 1 / F_s$, which has been encoded with random complex data. The white curves reproduce the power spectra from the middle column. For all power spectra a possibly existent discrete carrier line is omitted. (**a**) A single NRZ impulse "shaped" by a one-tap "filter" of order $R = 0$ leads to a sinc-shaped spectrum. (**b**) An ideal sinc-impulse is truncated by a rectangular window. The corresponding filter is of order $R = 16$. The power spectrum results from the convolution of the rect-shaped spectrum of the sinc-impulse with the sinc-shaped spectrum of the rect-window. (**c**) An increased filter order of $R = 32$ leads to a larger time window, and therefore the resulting spectrum evolves towards an ideal rectangular shape. (**d**) A filter with very high order $R = 1024$ closely approximates a rect-shaped power spectrum. Overshoots and ringing are due to Gibbs' phenomenon. All pulses in these plots have been q-fold oversampled with $q = 2$.

Nyquist signals shaped with various FIR filters are depicted in Fig. 4.6(b)–(d). The filter order R with $R + 1$ taps corresponds to the rectangular time window within which the function is defined (left column, green). The convolution of the rectangular spectrum of an infinitely extended temporal sinc-pulse with the sinc-shaped spectrum of the rectangular time window leads to the power spectra depicted in Fig. 4.6(b)–(d), middle and right column. As the filter order R increases from $R = 16$ to $R = 1024$, the spectrum evolves towards an ideal rectangle rect(f / F_s) with a spectral width equal to the Nyquist bandwidth F_s for complex data. Already for $R = 32$ a significant increase of the spectral efficiency is to be seen in comparison to NRZ modulation. For $R = 1024$ the ideal rectangular spectrum is approximated even more closely. However, due to Gibbs' phenomenon, strong ringing at the steep spectral slopes is to be observed. Non-rectangular window functions like Hann or Hamming windows lead to smoothened spectra and a stronger suppression of the side lobes. However, this advantage comes at the price of a widened spectrum and thus a reduced spectral efficiency.

4.1.4 Spectral efficiency and peak-to-average power ratio

Spectral efficiency (SE) is a major argument for the use of advanced modulation formats in combination with sophisticated multiplexing techniques. Since Nyquist pulses and OFDM signals are closely related, it is interesting to compare the potential SE of both techniques. To this end we compute the spectral width B of the Nyquist pulse up to the first zero outside the main band, which has a width F_s. The same definition is also used for OFDM [23]. The SE results from relating the information rate F_d (measured in bit/s) to the required transmission bandwidth B, SE $= F_d / B$. Information rate and symbol rate are related as follows: For M-ary single-polarization single-carrier Nyquist pulse transmission, the symbol rate is $F_s^{\mathrm{Nyq}} = F_d / \log_2 M$ (in the Nyquist context abbreviated by $F_s = F_s^{\mathrm{Nyq}}$, see Table A.2). For single-polarization M-ary OFDM signals with N subcarriers the symbol rate amounts to $F_s^{\mathrm{OFDM}} = F_d / (N \log_2 M)$ (abbreviated in the OFDM context by the same symbol $F_s = F_s^{\mathrm{OFDM}}$, see Table A.2).

The transmission bandwidth depends on the respective modulation types and formats. For Nyquist pulses, the spectrum is calculated in the Appendix, Eq. (6.2.26). Because of the finite length of the actual Nyquist pulses, the spectrum depends on the filter order R and the oversampling factor q, see Fig. 4.6. For convenience and without loss of generality we choose the spectral symbol $i = 0$ which lies symmetrical to $f = 0$. The spectrum then reads

$$Y^{(0)}(f, R) = \frac{T_s}{\pi}\left[\mathrm{Si}\left(\pi R \frac{f + F_s/2}{qF_s}\right) - \mathrm{Si}\left(\pi R \frac{f - F_s/2}{qF_s}\right)\right]. \tag{2.5.5}$$

The function Si(z) denotes the sine integral [124], see text before Eq. (6.2.26) in the Appendix. Power spectra computed from Eq. (2.5.5) closely match the graphs of Fig. 4.6 which are obtained by simulations. To determine the bandwidth $B = B^{\mathrm{Nyq}}$, we find the first spectral zeros to the right and to the left of the main band by a numerically exact evaluation of Eq. (2.5.5). From these results we extract a simple empirical relation to estimate the SE of digitally generated Nyquist signals:

$$\mathrm{SE}_{\mathrm{Nyquist}} = \begin{cases} \dfrac{\log_2 M}{1+2.517/R} & \text{for} \quad 1 \le R \le 1024, \ q=2 \\ \log_2 M & \text{for} \quad R > 1024, \ q=2 \end{cases} \tag{2.5.6}$$

The resulting spectral efficiency according to Eq. (2.5.6) is plotted in Fig. 4.7(a) (blue line).

For OFDM the SE is influenced by the number of subcarriers N, or in other words by the size of the inverse fast Fourier transform (IFFT) used for signal generation. For our discussion we disregard more advanced OFDM techniques such as a cyclic prefix, guard bands or the introduction of pilot tones that would decrease the SE. The resulting SE then is [23]

$$\mathrm{SE}_{\mathrm{OFDM}} = \frac{\log_2 M}{1+1/N}. \tag{2.5.7}$$

The normalized spectral efficiencies of OFDM signals are also depicted in Fig. 4.7(a) (red line). The SE of both techniques is almost equal.

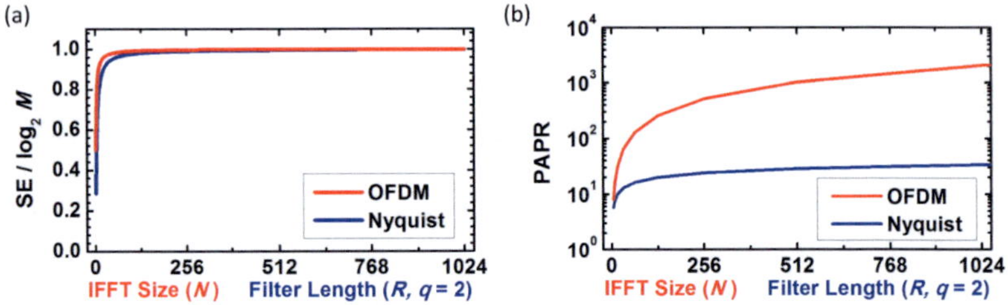

Fig. 4.7 Spectral efficiency (SE) and peak-to-average power ratio (PAPR) for Nyquist pulses and OFDM signals as a function of the number R of filter taps and the IFFT size N, respectively. (**a**) Normalized spectral efficiency (SE) for single-polarization M-ary QAM modulation. Nyquist shaped signals and OFDM signals show almost identical SE. (**b**) PAPR increases with filter order R for single-carrier Nyquist signals slower than the PAPR for OFDM with the number of subcarriers N.

A major issue that is often referred to reporting on OFDM is the high peak to average power ratio (PAPR) of the time domain signal. This is due to the coherent superposition of multiple sinusoidal carriers that could interfere constructively. As a consequence, high signal amplitudes can occur. In the following we derive PAPR expressions at the transmitter side for Nyquist pulse transmission and for OFDM signaling. At the transmitter, a large PAPR is most critical regarding the rather low resolution of high-speed DACs, the conversion range of which has to be utilized optimally. The PAPR at the receiver end depends heavily on properties of the transmission link like dispersion or nonlinearity tolerance. Therefore, general predictions cannot be made.

To derive an expression for the PAPR in OFDM we need to find the peak power and an expression for the average power. A maximum value can be found as follows: In order to compute the largest possible peak power of an OFDM signal $x(t)$, we assume without loss of generality that the N subcarriers are modulated with a random sequence of real coefficients $c_{ik} = \pm 1$. In this case the maximum amplitude is seen if all N maxima of the temporal sinusoidals happen to add constructively at one point in time, see Fig. 4.3(a) at $t = 0$ and Eq. (6.2.21) in the Appendix. The average power of such a random OFDM signal is the sum of the average powers of the N orthogonal subcarriers. For arbitrary modulation coefficients c_{ik}, the average power for $c_{ik} = \pm 1$ has to be divided by a format dependent factor k^2 [81][82]. In real-world OFDM systems the infinitely extended ideal spectrum is narrowed by low-pass filtering, so that the orthogonality relation does not hold any more in the strict sense. Nevertheless, with the orthogonality relation Eq. (6.2.2) and the power relation Eq. (6.2.8) we obtain a good approximation of the average power given by Eq. (6.2.24) with $Q = N$. We thus approximate the $\mathrm{PAPR}_{\mathrm{OFDM}}$ by

$$\mathrm{PAPR}_{\mathrm{OFDM}} = \frac{N^2}{\frac{1}{2} N / k^2} = 2k^2 N. \tag{2.5.8}$$

The result of Eq. (2.5.8) for $k^2 = 1$ is seen in Fig. 4.7(b), red line. With increasing number N of subcarriers the value for $\mathrm{PAPR}_{\mathrm{OFDM}}$ increases linearly. However, the probability that an OFDM signal actually has this peak amplitude decreases with the complexity of the M-ary QAM modulation and with the number N of the subcarriers.

For Nyquist signals the PAPR has to be investigated, too, since a superposition of temporally shifted sinc-pulses, see Fig. 4.3(b), also produces high signal amplitudes at certain times. We assume again that the Nyquist pulses are modulated with a random sequence of real coefficients $c_{ik} = \pm 1$. Although the local extrema of a single sinc-impulse are not located at times $t / T_s = -0.5, 0.5, 1.5\ldots$, i.e., not in the center of the interval between zeros, it can be shown that the extrema of superimposed Nyquist pulses are located at exactly these times, Eq. (6.2.12) in the Appendix. For a worst-case consideration all contributions sum up constructively, so in order to obtain the maximum compound signal we sum up the absolute values of sinc-pulses at $t / T_s = 1 / 2$. If the compound Nyquist signal was constructed with infinitely extended sinc-functions, the maximum signal power would not converge when the number of Nyquist pulses increases. Nevertheless, sinc-functions located far away from the time of summation only contribute little to the sum. For a finite approximation of a sinc-impulse as described in Section 4.1.3, only R / q pulses can contribute. Here the filter order R denotes the number of time intervals T_s / q for q-fold oversampling, i.e., R stands for the length of the impulse response. We find the maximum power, see Eq. (6.2.13) in the Appendix with $Q = R / q$

$$P_{\max} = \left[\sum_{r=-R/(2q)+1}^{R/(2q)} \left| \operatorname{sinc}\left(\frac{1}{2} - r \right) \right| \right]^2 . \tag{2.5.9}$$

Technically speaking, $P_{\max}$ could become arbitrarily large for large filter orders R. Yet, while $P_{\max}$ increases with R, the probability for finding R sinc-pulses interfering constructively decreases as well similarly to the OFDM case.

For finalizing the calculation of the PAPR, we need the average power of a single-carrier Nyquist signal $y^{(0)}(t)$ encoded with real coefficients $c_{ik} = \pm 1$. According to [125] we find the average power $\bar{P}$ of an ideal Nyquist signal (see Eq. (6.2.9) in the Appendix),

$$\bar{P} = \frac{1}{T_s} \int_{-\infty}^{+\infty} \operatorname{sinc}\left(\frac{t}{T_s}\right) \mathrm{d}t = 1. \tag{2.5.10}$$

As for band-limited OFDM spectra, orthogonality is lost for truncated Nyquist sinc-impulses. If Nyquist pulses are generated with a filter of finite (but sufficiently large) order R, orthogonality as implied by Eq. (2.5.10) is still a good assumption, so that the average power of truncated Nyquist sinc-impulses is close to $\bar{P} = 1$. As before, for arbitrary modulation coefficients c_{ik}, Eq. (2.5.10) has to be divided by a format dependent factor k^2 [81][82]. The $\mathrm{PAPR}_{\mathrm{Nyquist}}$ then follows from the ratio of maximum power P_{max} and average power $\bar{P} \approx 1/k^2$,

$$\mathrm{PAPR}_{\mathrm{Nyquist}} = \frac{P_{\max}}{\bar{P}} \approx k^2 \left[\sum_{r=-R/(2q)+1}^{R/(2q)} \left| \operatorname{sinc}\left(\frac{1}{2} - r\right) \right| \right]^2 . \tag{2.5.11}$$

A detailed mathematical description is given in the Appendix, leading to Eq. (6.2.19). The PAPR of Nyquist signals from Eq. (2.5.11) and $k^2 = 1$ is plotted in Fig. 4.7(b), blue line. Unlike OFDM signals where the PAPR increases linearly, Eq. (2.5.8), the PAPR of Nyquist signals does not, due to the temporal decay of its elementary sinc-impulse. However, neither for OFDM nor for single-carrier Nyquist pulses the PAPR converges with increasing IFFT size N or filter order R, respectively.

In Nyquist WDM systems, the PAPR could be higher, because multiple spectral symbols separated by at least F_s might add up constructively as well. However, the Nyquist channel spacing is typically in the order of several GHz [31][121][19] whereas OFDM carrier spacings are often chosen to be in the MHz [126][39] range. For large channel spacings as in Nyquist WDM, however, strong signal peaks only occur for very short times, and dispersion causes signal peaks to decay rapidly if large frequency differences are involved. Non-linear effects in WDM systems have been investigated in [127].

4.2 Real-Time Nyquist Pulse Transmitters

In the following section, the design and implementation of real-time Nyquist pulse-shaping is discussed. Therefore FPGAs are employed to handle data rates beyond 100 Gbit/s.

4.2.1 Efficient Real-Time Nyquist Tx Using Look-Up Tables

Real-time Nyquist pulse generation beyond 100 Gbit/s and its relation to OFDM

R. Schmogrow, M. Winter, M. Meyer, A. Ludwig, D. Hillerkuss, B. Nebendahl, S. Ben-Ezra, J. Meyer, M. Dreschmann, M. Huebner, J. Becker, C. Koos, W. Freude, and J. Leuthold

Opt. Express 20(1), pp. 317–337 (2012). [J22]

4.2.1.1 Implementation

In order to electronically generate Nyquist pulses, advanced digital signal processing (DSP) along with digital-to-analog converters are needed. Suitable devices for high-end DSP are either application-specific integrated circuits (ASIC) or field programmable gate arrays (FPGA). Since ASIC development is time consuming and comes along with high financial efforts, the use of FPGA for prototyping purposes is well established.

FPGA based DSP: The main challenge for real-time Nyquist pulse generation is the development of high performance FIR filters that provide a sufficient number of filter taps, have adequate precision, and enable a high data throughput. The extensive use of look-up tables (LUT) is a highly efficient way to implement FIR filters on FPGAs, since resource hungry complex multiplications can be avoided. The principle scheme of the DSP performed by the FPGA is depicted in Fig. 4.8. Due to the strong parallelization of the processing, the FPGA internal clock can be significantly lower than the Nyquist pulse rate. The FPGA produces 128 samples (each of 6 bit depth) per clock cycle (4.57 ns for 28 GSa/s). In each computation window (and for an oversampling factor of $q = 2$), a series of $128 / 2 = 64$ modulated sinc-pulses form the output signal, Fig. 4.8. The FIR filter is realized in the time domain by convolving the complex modulation coefficients c_{ik} with a sampled sinc-shaped impulse. All possible products of c_{ik} and an elementary sinc-impulse are sampled (dots in Fig. 4.8(a)), quantized and stored within LUTs. For illustration purposes we choose $c_{ik} = \pm 1$ as coefficients for the spectral symbol $i = 0$. The LUT outputs are delayed by a multiple of T_s, Fig. 4.8(b), and all samples belonging to the same point in time are added, Fig. 4.8(c). The resulting output is fed to a clipping module, Fig. 4.8(d), which then delivers the output Nyquist waveform as seen in Fig. 4.8(e). Red dots mark the position of the sinc-pulse maxima ±1. Continuous operation for an infinite number of Nyquist pulses is achieved through cyclic buffering of samples that are used within adjacent computation windows.

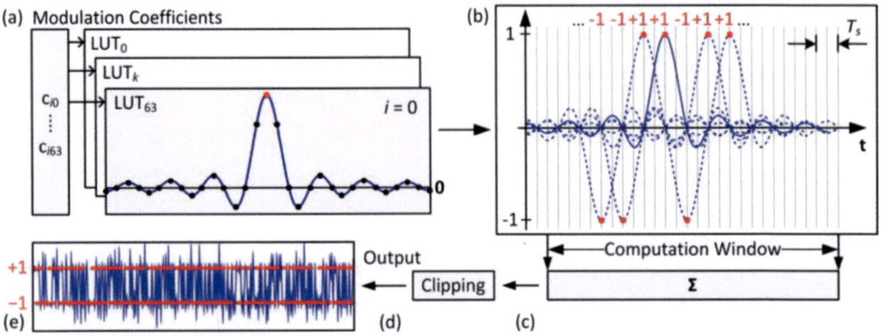

Fig. 4.8 FPGA parallel processing. In each computation window the FPGA produces a series of 64 sinc-pulses with different weighting coefficients $c_{ik} = \pm 1$ and for $i = 0$. (**a**) Products of modulation coefficients and windowed sinc-pulses (two-fold oversampled, $q = 2$) are stored within 64 lookup tables (LUT). Stored samples within the LUT are marked as dots. Each of the 64 LUT outputs is delayed in time such that (**b**) sinc-maxima (red dots) fall on the zeros of neighboring sinc-functions corresponding to a spacing by T_s. For each of the 128 sampling times inside the computation window the non-zero samples are added (**c**), clipped (**d**), and eventually produce the output waveform (**e**).

Experimental setup: Our real-time Nyquist pulse transmitter (Tx) comprises two Xilinx Virtex 5 field programmable gate arrays (FPGA), two high-speed Micram DACs with 6 bit resolution, a nested $LiNbO_3$ Mach-Zehnder modulator (MZM) serving as an I/Q-modulator, and an erbium-doped fiber amplifier (EDFA). We modulate a continuous wave (CW) external cavity laser (ECL) with in-phase (I) and quadrature-phase (Q) data as shown in Fig. 4.9. Within the FPGAs, complex Nyquist pulses are calculated from a $2^{15} - 1$ PRBS in real-time as outlined above, and passed on to the DACs. The polarization division multiplexing (PDM) stage then emulates a polarization multiplexed signal [78].

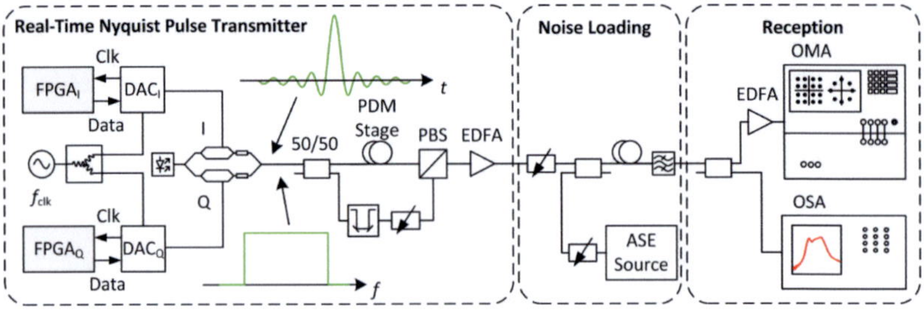

Fig. 4.9 Experimental setup for a real-time Nyquist pulse transmitter comprising $FPGA_{I,Q}$ and $DAC_{I,Q}$, an externally coupled laser (ECL), an optical I/Q-modulator and an EDFA. An "ideal" elementary output pulse along with its corresponding spectrum is shown as an inset (green). The PDM stage emulates polarization division multiplexing. The signal is loaded with noise by adding a variable amount of amplified spontaneous emission (ASE). The receiver comprises an EDFA along with an Agilent N4391A optical modulation analyzer (OMA) for BER and EVM measurements, and an optical spectrum analyzer (OSA) for OSNR measurements.

We use two complex samples in each symbol time slot leading to a symbol rate of 14 GBd for 28 GSa/s operating DACs. The resulting oversampling by a factor of $q = 2$ can be reduced if adequate electrical or optical anti-aliasing filters are available.

An amplified spontaneous emission (ASE) source adds optical noise to the signal. The noise power can be adjusted by a variable optical attenuator (VOA). A second VOA is used to additionally attenuate the signal when measuring very low optical signal-to-noise ratios (OSNR). An optical band-pass filter removes noise components outside the signal spectrum.

At the receiver (Rx) the signal is split in two. One part feeds an optical spectrum analyzer (OSA) that measures OSNR values for different levels of noise loading. The other part is amplified by an EDFA before it is detected by an Agilent optical modulation analyzer (OMA). The OMA performs the offline processing including carrier phase and clock recovery as well as decoding, bit error ratio (BER), and error vector magnitude (EVM) measurements.

Experimental results: We performed measurements with PDM-QPSK and PDM-16QAM signals pre-shaped by FIR filters with order 16 and 32 as well as raw NRZ for various levels of OSNR. Measured spectra (colored, noisy curves) for both filters and raw NRZ (black) are seen in Fig. 4.10(a). The results are consistent with simulations of the FPGA's VHDL code using the ModelSim software (white lines). As expected, the signals generated with $R = 32$ showed best spectral efficiency and a nearly rectangular shape. The noise floor outside the main band is mostly due to quantization noise (6 bit DAC resolution). Quantization noise does not significantly degrade the signal quality when the transmitter is employed in a Nyquist WDM system [31], but choosing too small a filter order R deteriorates the signal severely. The displayed constellation diagrams are constructed by overlaying the diagrams for both polarizations. The results for both filter orders and NRZ are depicted in Fig. 4.10(b). In order to evaluate the signal quality, the OMA measures BER (symbols in Fig. 4.10(c)) and EVM. The EVM values are converted to a BER estimate (dashed lines in Fig. 4.10(c)) according to [81][82]. The format-dependent factor k, see Eq. (2.5.8), and (2.5.11), converts an EVM defined by the outermost constellation point (EVM_m) to an EVM defined by the average power (EVM_a) [81][82]. EVM and OSNR are approximately related by $k \times \mathrm{EVM}_m \approx 1/\sqrt{\mathrm{OSNR}}$ [81][82]. The measured function BER(OSNR) (symbols, lower horizontal axis) and the estimated function BER(EVM) (dashed lines, upper horizontal axis) are plotted in Fig. 4.10(c). The graph shows good agreement between measured BER and estimated BER over a broad range. We see no degradation of the signal quality comparing the pulse sequences generated by filters with order 16 and 32 and NRZ. The spectral efficiency for PDM-16QAM ($R = 32$) is 7.5 bit/s/Hz compared to the theoretical limit of 8 bit/s/Hz.

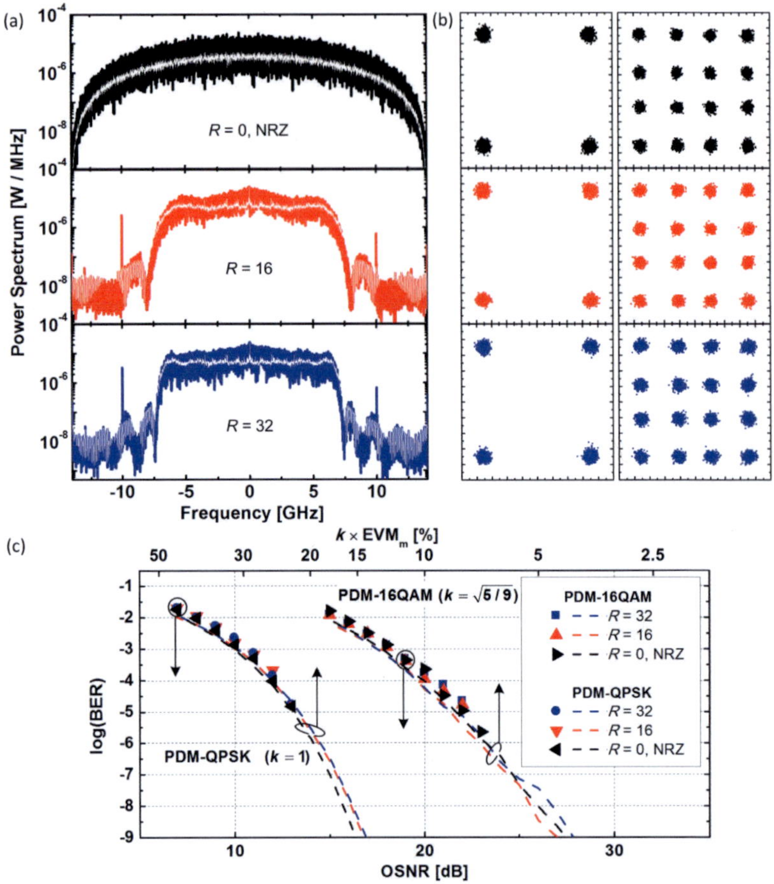

Fig. 4.10 Spectra, constellation diagrams and BER versus OSNR and EVM results for PDM-16QAM and PDM-QPSK signals generated with $R = 0$ (black, raw NRZ), $R = 16$ (red), and $R = 32$ (blue) filters. Plotted constellation diagrams comprise data from both polarizations. (**a**) Ensemble averages of measured (black, red, and blue) and VHDL simulated (white) spectra. It may be seen that with an increasing number of taps the spectrum evolves towards an ideal rectangle. The observed noise floors next to the spectra are due to quantization noise and do not significantly degrade the signal quality. (**b**) Constellation diagrams for PDM-QPSK and PDM-16QAM at highest possible OSNR. (**c**) Measured BER (symbols) versus OSNR (lower axis). Calculated equivalent BER derived from measured EVM (dashed lines, upper axis) are plotted for Nyquist signals generated with filter of order 16 and 32 and raw NRZ. Calculated EVM values corresponding to the measured OSNR are specified on the upper axis.

4.2.2 Real-Time Nyquist Signaling with Dynamic Precision and Adaptable Signal Bandwidth

Real-time Nyquist signaling with dynamic precision and flexible non-integer oversampling

R. Schmogrow, M. Meyer, P.C. Schindler, M. Dreschmann, J. Meyer, S. Ben-Ezra, J. Becker, C. Koos, W. Freude, and J. Leuthold

Opt. Express 22(1), pp. 193–209 (2014). [J1]

4.2.2.1 Introduction

Next generation networks rely on real-time transmitters (Tx) and receivers (Rx) that allow the generation and reception of phase and amplitude modulated signals in combination with spectrally efficient pulse-shapes. However, real-time processing of multi-gigabit data streams is highly demanding and only practical if algorithms are found that provide high computational precision with little processing effort. In addition, symbol rates should be as close as possible to the sampling rates offered by state-of-the-art digital-to-analog and analog-to-digital converters (DAC and ADC) in order to efficiently make use of the available speed on the one hand and for minimum energy consumption on the other hand. Last but not least, developed techniques should be compatible with emerging novel flexgrid networks by offering dynamic adaptation of symbol rate and bandwidth.

In view of signals with optimum spectral efficiency (SE), there are two main contenders, namely Nyquist signaling [24] and orthogonal frequency division multiplexing (OFDM) [23]. Both pulse shaping techniques offer a SE close to the theoretical limit. Real-time Tx and Rx for up to 101.5 Gbit/s (Tx) [38][39][79] and up to 41.25 Gbit/s (Rx) [35][36][128] have been already demonstrated. While these numbers are already impressive, the ultimate speed is limited by the specific implementation needs for either pulse-shaping technique. Recently, the limitations for digital real-time Nyquist and OFDM systems have been investigated [129]. While Nyquist signaling offers advantages such as higher out-of-band suppression [129] or a potentially lower peak-to-average power ratio [79], it also poses implementation challenges: Nyquist pulse-shaping requires a high computational precision due to the fast decay of the sinc-shaped impulse responses, and a large sampling rate with the usually used two-fold oversampling. Finally, processing of data rates beyond 100 Gbit/s can be only achieved using a high degree of parallelization, which is a challenge of its own.

Another prominent requirement for the application of digitally processed Nyquist and OFDM signals in flexgrid networks is the capability to change the symbol rate and thereby the signal bandwidth during runtime. In order to elaborate the challenges of optical flexgrid networks, Fig. 4.11 exemplarily shows a network with three frequency channels

(f_0 red, f_1 black and f_2 blue). We illustrate how input data are processed by digital signal processing (DSP) and fed to two DACs. The figure shows spectra of signals at different positions in the Tx as insets. For the signal measured at the DAC outputs, it can be seen that the spectra comprise both, the spectrum of interest (black) and also so-called image spectra (white) that repeat infinitely for an ideal DAC. The latter are removed by electrical low-pass filters (transfer function: dashed line). Nyquist sampling at the symbol rate would call for non-realizable "brick wall" filters, therefore some degree of oversampling by a factor q is required for accommodating real-world filter slopes in the guard bands (GB) between spectrum of interest and adjacent images. The oversampled data are then encoded on an optical carrier (laser diode, LD) with an optical modulator. Subsequently, the signals of several Tx (framed red, black, and blue) are multiplexed (MUX). The Rx then coherently receives one of the demultiplexed (DEMUX) channels using a 90° optical hybrid and photo detectors followed by another set of analog electrical filters, ADCs, and DSP.

The ultimate versatility is obtained when Tx and Rx can be flexibly adapted to optimally utilize the assigned bandwidth in the network (Fig. 4.11, Scenarios A and B). Unfortunately, the DSPs and filters do not allow for such flexible adaptation of the bandwidth by changing the sampling frequency. Designs for DSP circuits (e.g. field programmable gate arrays, FPGA, or application specific integrated circuits, ASIC) are typically built to be operated at a fixed frequency only. Any major change of the clock rate would require either a reprogramming or even a redesign. Likewise, the cut-off frequency of the analog image rejection (Tx) and anti-alias (Rx) filters is fixed by design and therefore removing any image spectra (Tx) or preventing aliasing (Rx) is done for one fixed sampling frequency only.

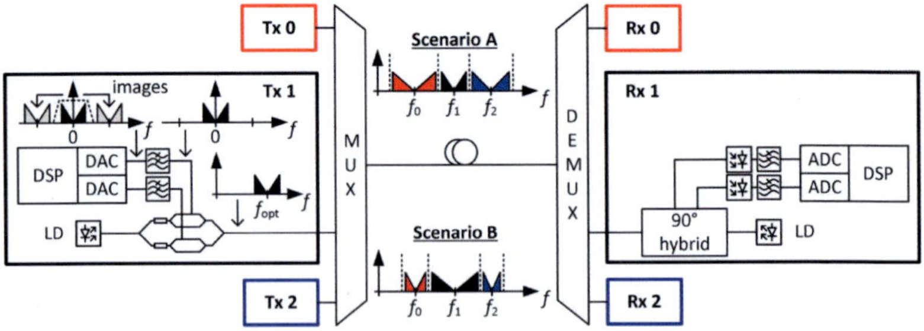

Fig. 4.11 Vision of a flexgrid optical network, where various Tx generate signals with adjustable bandwidths. In this example, the middle channel Tx (at frequency f_1) adapts its signal bandwidth flexibly. In Scenario **A**, Tx 1 generates signals with small bandwidth as its neighboring Tx 0 and Tx 2 (f_0 and f_2) occupy large bandwidths. In Scenario **B,** Tx 1 may use more bandwidth as its neighbors occupy small bandwidths. To realize networks with channels of flexible bandwidths, advanced algorithms for Tx and Rx digital signal processing (DSP) need to be found especially as the hardware does not support changing clock frequencies.

Alternative ways to change the bandwidth without changing the sampling frequency need to be found. This is possible by tuning the oversampling factor q rather than the sampling frequency (rate) f_s of the signal converters (DACs and ADCs). To understand this process, we take a closer look at oversampling. According to the Nyquist sampling theorem a bandwidth-limited signal can be fully represented by two real-valued or one complex sample per period of the highest frequency component in the signal [61]. This corresponds to one complex sample per symbol period, i.e., $q = 1$. In general, the relation between symbol rate F_s, sampling rate f_s, and oversampling factor q follows from:

$$\textit{symbol rate}\, F_s = \frac{\textit{sampling rate}\, f_s}{\textit{oversampling factor}\, q}. \tag{2.5.12}$$

For Nyquist sinc-pulses, the signal bandwidth B equals the symbol rate $B = F_s$. For Nyquist signals comprising so-called "raised-cosine" pulses [77], the relation between symbol rate and signal bandwidth is defined through the roll-off factor β and is described by $B = F_s\,(1 + \beta)$ for $0 \leq \beta \leq 1$. In general, changing the symbol rate F_s and thus the signal bandwidth B causes a change of the oversampling factor q if the sampling rate f_s is kept constant.

The effect of changing the symbol rate F_s (the bandwidth B) with constant sampling rate f_s is illustrated in Fig. 4.12. It shows both, Nyquist signals and OFDM signals, processed with different q but constant f_s. If signals are not oversampled, i.e., $q = 1$, then main spectrum and image spectra adjoin, see Fig. 4.12(a) and 12(b). Ideal "brick wall" analog filters would be needed to fully remove the image spectra without affecting the main spectrum for Nyquist signals, Fig. 4.12(a). For OFDM signals with $q = 1$ it is impossible to filter the images without affecting the main spectrum. In order to remove the image spectra with realizable filters, a guard band (GB) is needed. This is obtained for an oversampling factor $q > 1$, see Fig. 4.12(c)–12(f) for Nyquist and OFDM signals with $q = 2$, and $q = 4/3$, respectively [63]. In practice, an oversampling factor $q = 2$ is commonly used [79], because in this case the DSP implementation is straight-forward. It can be seen in Fig. 4.12(c)–12(f) that for a fixed sampling frequency f_s, the same analog filters as for $q = 2$ (schematic transfer function indicated by dashed lines) can be used to remove the image spectra. Therefore the signal bandwidth can be changed through the oversampling factor q, while the hardware and sampling frequency f_s is kept constant. In OFDM systems, the signal bandwidth and thus the oversampling factor q can be simply changed by nulling more or less of the SCs close to the edge of the main band. For Nyquist signals that are usually generated using finite-duration impulse response (FIR) filters, effectively changing the signal bandwidth through an adaptive oversampling factor q is more challenging.

In this paper, we first discuss a novel real-time sinc-pulse shaping algorithm that provides superior signal quality compared to conventional fixed-point arithmetic, while keeping the processing effort at minimum. This is obtained by dynamically changing the effective word length for computation [41]. As a result, we can decrease the bit error ratio (BER) significantly without increasing the required computational resources. Second, we reduce the usual oversampling factor $q = 2$ to $q = 4/3 \approx 1.33$, thus increasing the symbol rate F_s at a constant sampling frequency f_s by 50 % [44]. Third, we present an efficient parallel processing technique for generating and receiving Nyquist signals with adaptable oversampling factor q in order to process signals with variable symbol rates and bandwidths. All presented results are experimentally validated using a real-time Tx. To verify the advantages of processing with dynamical computational precision, we transmit polarization division multiplexed (PDM) 16QAM (64QAM) signals over up to 300 km (150 km) of standard single mode fiber (SSMF). To demonstrate Nyquist signals with a rational oversampling factor of $q \approx 1.33$, we transmit QPSK, 16QAM, and PDM-64QAM signals. A maximum bit rate of 252 Gbit/s (PDM-64QAM) is achieved for signals transmitted over up to 100 km of ultra-large area fiber (ULAF). Although showing the Tx only, the developed algorithms are also suitable for a potential real-time Rx.

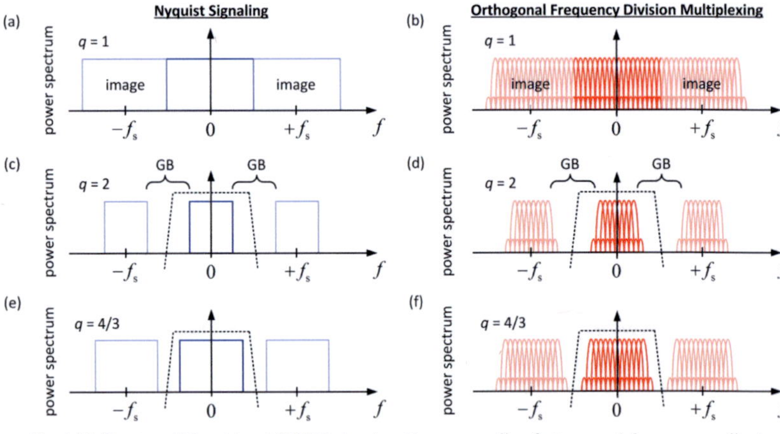

Fig. 4.12 Spectra of Nyquist and OFDM signals with oversampling factors $q = 1$ (no oversampling), $q = 2$ and $q = 4/3$. The main spectra are centered on zero frequency. Image spectra are displayed in light color at multiples of the sampling frequency f_s. These spectra are removed by analog filters (schematic transfer function indicated by dashed lines). **(a)** Spectrum of a Nyquist signal without oversampling. **(b)** OFDM spectrum ($q = 1$). **(c)** Electrical spectrum of a Nyquist signal ($q = 2$). The spectral guard band (GB) equals the electrical signal bandwidth. **(d)** OFDM spectrum ($q = 2$). **(e)** Nyquist spectrum with reduced oversampling factor $q = 4/3$. For a fixed sampling frequency f_s the symbol rate is increased while the GB is reduced. **(f)** Corresponding OFDM spectrum for $q = 4/3$.

4.2.2.2 Nyquist DSP with dynamically adjusted precision

Generating Nyquist signals with maximum SE calls for FIR filters of high order R [79]. However, increasing the filter order R (or in other words the number of filter coefficients) is problematic for sinc-pulses, since coefficients representing the filter's impulse response (IR) far away from the center peak of the IR show very small magnitudes.

On the one hand, high computational precision is needed to correctly represent small and large magnitudes of the IR at the same time. On the other hand, FPGAs and ASICs are the only reasonable choices for multi-gigabit processing, and they typically use fixed-point integer arithmetic so that the computational effort scales with the word length.

To provide a large effective word length without increasing the processing effort we dynamically adapt the computational precision within different intervals of the IR. To do so, an elementary sinc-impulse (with amplitude '1' at $t = 0$) is divided into intervals that share a multiplication factor 2^s, Fig. 4.13(a). All sampled floating-point (FP) values (word length 32 bit) of the sinc-impulse are scaled to a maximum signed value of $2^5 - 1$, which is chosen to match the final DAC resolution of 6 bit. The outer sinc-magnitudes (far away from $t = 0$) are close to zero. Therefore, each FP value is multiplied by 2^s, where the weight s is chosen — according to the sinc-magnitude — such that rounding to the closest 6 bit signed integer results in a maximum number of non-zero most significant bits.

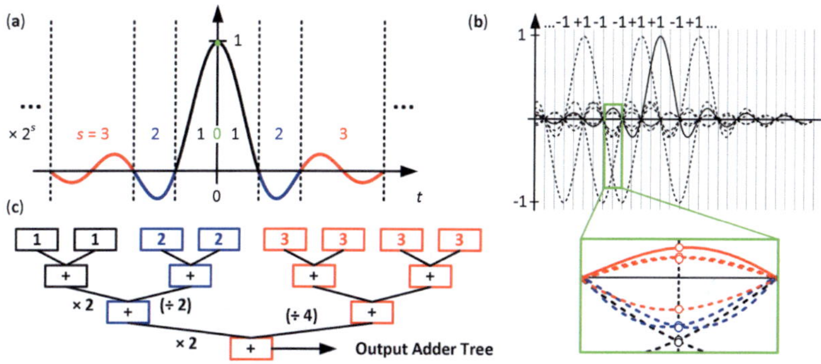

Fig. 4.13 DSP with dynamical increase of the computational word length. (**a**) An elementary sinc-impulse with amplitude 1 at $t = 0$ is scaled, multiplied with factors 2^s, and rounded to a signed 6 bit integer. The weight s is chosen according to the magnitude decrease of the sinc-function. (**b**) An output waveform resulting from the superposition of various sinc-impulses. Samples (see green-framed blow-up) for each point of time are added. (**c**) Adder tree for groups with different s uses either multiplications (×, more accurate) or divisions (÷) to merge the groups.

The output signal is constructed from the superposition of modulated elementary sinc-impulses. A train of pulses is shown in Fig. 4.13(b). For simplicity, the modulation coefficients are chosen to be ± 1. The output waveform is computed by adding the samples of all impulses belonging to the same point in time. The inset (framed green) shows 8 samples (○, ○, ○) to be added. An adder tree as seen in Fig. 4.13(c) groups samples with equal s and computes the output values, taking into account the previously assigned weights. When adding groups with unequal weights (e.g. $s_1 = 1$ and $s_2 = 2$ as in Fig. 4.13(c)), either the values of group s_1 must be divided by 2^{s1-s2} (×2 in Fig. 4.13(c)), or the values of group s_2 must be divided by 2^{s2-s1} (÷2 in Fig. 4.13(c)). The (×)-procedure results in better accuracy than the (÷)-method. This algorithm has to be applied for all adder stages and results in an improved effective word length of $p_{\text{eff}} = 9.9$ bit, instead of 7.9 bit for conventional, "standard" DSP.

4.2.2.3 Nyquist DSP with rational oversampling factor

Digital-to-analog converters and ADCs only offer sampling rates f_s up to a certain speed. This sets an upper limit to the achievable symbol rate F_s. If an oversampling factor $q = 2$ is used as in previous work [79], then the achievable symbol rate F_s is limited to $f_s / 2$. To overcome this limitation, we introduce a rational oversampling factor $q = k / l$ and discuss how such a filter can be implemented without computational overhead.

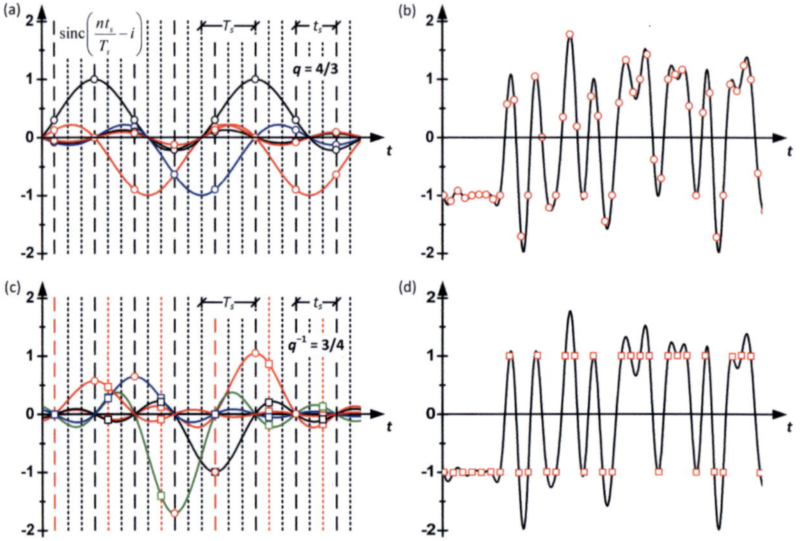

Fig. 4.14 Nyquist signaling with rational oversampling factor $q = k / l = 4 / 3$. The sampling period t_s is a fraction of T_s. (**a**) At the Tx, only every l-th sample is considered (open circles, dashed grid) instead of computing k samples per symbol (dotted grid). The IR is sampled at different relative sampling positions which repeat after $l = 3$ pulses (black, red, and blue). (**b**) Output waveform at the Tx. (**c**) At the Rx, the sampled signal (open circles) is interpolated with sinc-functions to recover the transmitted data. This time, only every k-th sample needs to be processed (open squares, red gridlines). A number of $k = 4$ differently sampled sinc-functions (black, red, blue, and green) need to be provided as IR of the Rx filter. (**d**) Waveform after resampling. Data is recovered ISI-free (open squares).

One way to generate sinc-pulses with an arbitrary rational $q > 1$ is to operate the FIR pulse-shaper at a clock rate $k\,F_s$ and to use every l-th sample for defining the output waveform while dropping all samples in between. This method, however, is unfavorable in terms of computational complexity, as samples have to be processed that do not contribute to the output. The same argument holds for the Rx, where $q = k / l$ oversampled signals need to be resampled by $q^{-1} = l / k$ so that at the output of the Rx FIR filter, exactly one sample per transmitted symbol, is obtained. In the following, we discuss a method to directly process q-fold oversampled sinc-pulses without overhead as will be explained for the example $q = 4 / 3$.

The processing performed in the Tx is indicated in Fig. 4.14(a). It shows a number of sinc-pulses separated by the symbol period T_s. For the sake of visibility and without loss of generality, all sinc-pulses are modulated with real modulation coefficients ±1. Instead of computing all samples for ($k = 4$)-fold oversampling (indicated by the total of dotted and dashed grid lines), only samples located on the dashed grid (open circles) need to be summed to form the output waveform shown in Fig. 4.14(b). It should be noted, that since the sample period $t_s = 1 / f_s$ is a fraction of the symbol period T_s, we need to sample the IR at different relative positions. Fortunately, these relative sample positions repeat after l pulses so that only $l = 3$ differently sampled sinc-pulses (black, red and blue) are required for generating the output waveform in Fig. 4.14(b). The processing performed in the Rx is indicated in Fig. 4.14(c). The sampled signal (open circles) is interpolated using sinc-functions. This is achieved by a similar FIR filter as has been used in the Tx. This time, however, the signal is resampled by a factor $q^{-1} = l / k = 3 / 4$. Instead of computing all samples lying on the depicted grid (dotted and dashed) in Fig. 4.14(c), only every k-th sample needs to be considered (open squares, red grid-lines). Therefore, just as it is done in the Tx, the sinc-shaped IR of the FIR filter is sampled at different relative positions which repeat after $k = 4$ pulses (black, red, blue, and green). Summing all samples (open squares) on this grid, the transmitted data (here: modulation coefficients ±1) can be recovered without intersymbol interference (ISI). The outcome is depicted in Fig. 4.14(d) where the reconstructed waveform is shown.

The situation changes if neighboring Nyquist channels exist and the spectral guard band between adjacent channels is small. In this case, fragments of the neighbors are not removed by the analog electrical anti-alias filters preceding the ADCs, red-shaded spectral portions in Fig. 4.14(a). The appropriate Rx digital filtering for this scenario is visualized in Fig. 4.14(a). The analog signal is sampled by an ADC which, in this example, is operated at the same sampling rate as the Tx DACs. In general, the ADC's sampling rate may be chosen as low as 1 sample per symbol (SPS) without introducing any penalty as long as neighboring channels are fully removed in the analog domain. In this example, however, a digital representation of the signal with $4 / 3 \approx 1.33$ SPS is obtained. Looking at the spectrum to the right of the ADC (second from left in Fig. 4.14(a)), one identifies the band of interest (white rectangle) along with the fragments of the next neighbors (red shapes). After upsampling by a factor of 3, the signal is now available with 4 SPS, so that the Rx spectrum is repeated 2 more times (second from right in Fig. 4.14(a)). Interpolation with a sinc-function (blue dashed box) as described by the preceding paragraph would only remove the periodic repetitions (frequency response rect(f): blue dashed rectangle) but not the fragments of the neighboring channels. These would lead to crosstalk during the subsequent downsampling process. In order to remove these fragments, an additional filter (frequency response $H(f)$: red dashed rectangle) with IR $h(t)$ must be employed (red dashed box). Fortunately, the two filters can be combined to a single one having a frequency response rect(f) $H(f) = H(f)$, since rect(f) is wider than $H(f)$. So, instead of using an interpolation as shown in Fig. 4.14(c), we can directly implement the filter $H(f)$ with its impulse-responses $h(t)$ according to Fig. 4.15(b). It should be noted that unlike in Fig. 4.14(c), the zeros of the weighed IRs do not coincide with the pulse centers. Naturally, the filter coefficients for the procedure Fig. 4.14 differ from the ones for $h(t)$ in Fig. 4.15, otherwise, there is no structural difference and the Rx can be as efficiently implemented as described in Fig. 4.14.

If no neighboring channels are present, both methods shown in Fig. 4.14(c) and Fig. 4.15(b) lead to the same result shown in Fig. 4.14(d) and Fig. 4.15(c). Furthermore, the filter $H(f)$ does not need to be rectangular and can have a square root raised-cosine shape as well.

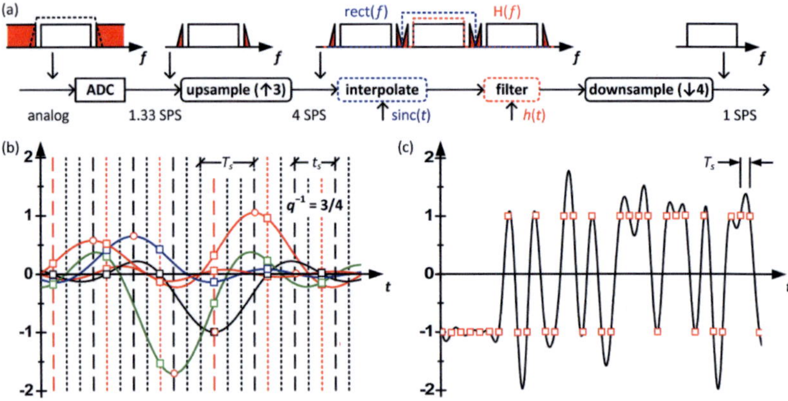

Fig. 4.15 Rx filter if neighboring channels are not fully removed by analog filters. (**a**) An ADC samples the analog signal with 1.33 samples per symbol (SPS). The spectrum shows fragments of the neighbors (red). Upsampling results in periodic repetitions. The interpolation filter (rect(f), blue dotted line) removes the periodic repetitions, but not the neighboring fragments (red). To avoid crosstalk during downsampling, these fragments have to be removed by an additional filter ($H(f)$, red dotted line). Both filters can be combined and replaced by rect(f) $H(f) = H(f)$. (**b**) Combined Rx resampling and filtering by using the same technique as described by Fig. 4.14(c). (**c**) Waveform after combined resampling and filtering results in the same signal as in Fig. 4.14(d).

We conclude that a significant amount of processing effort in Tx and Rx can be saved when filtering data with a rational q directly, instead of performing a conventional two-stage process (upsampling, interpolation, and downsampling) for rational oversampling factors. The technique is not limited to sinc-shaped Nyquist signaling, but can be equally well used to efficiently realize processing of signals with different pulse shapes (e.g. with a square root raised-cosine spectrum).

4.2.2.4 Parallel FIR filter design

To enable multi-gigabit processing in real-time, either ASICs or FPGAs are commonly used. While ASIC development is time consuming and costly, FPGAs are the best solution for prototyping. Employing FPGAs, multi-gigabit data streams can be processed at much reduced clock rates through parallelization. In this section we discuss a parallel FIR filter design suitable for FPGA and ASIC implementations. The design is made suitable for an adaptable rational oversampling factor q. This allows changing symbol rate and signal bandwidth without touching the sampling frequency.

Traditionally, FIR filters are described serially by a sequence of delay elements and $R + 1$ taps in-between. Signal samples $x[m]$ are weighed by the filter coefficients $h[r]$ and summed to form the output $y[m]$ [79] according to the discrete convolution

$$y[m] = \sum_{r=-R/2}^{+R/2} x[m-r]h[r], \qquad m,r \in Z \tag{2.5.13}$$

For a parallel realization of a filter described by Eq. (2.5.13) it should be first noted that each sample $x[m]$ at the filter input generates a weighed IR $x[m]$ $h[r]$ represented by $R + 1$ coefficients $h[r]$ at the filter output, where m, r represent discrete points in time. For a multiple of input samples $x[m]$, the output $y[m]$ follows from a linear superposition of all properly delayed IRs. Due to this linearity, we first generate all IRs $x[m]$ $h[r]$ individually as depicted in Fig. 4.16. If only a limited number of different input values $x[m]$ is possible (as it is true at the Tx for M-ary QAM signals), then the required multiplications can be pre-computed and the results stored in look-up tables (LUT) [79].

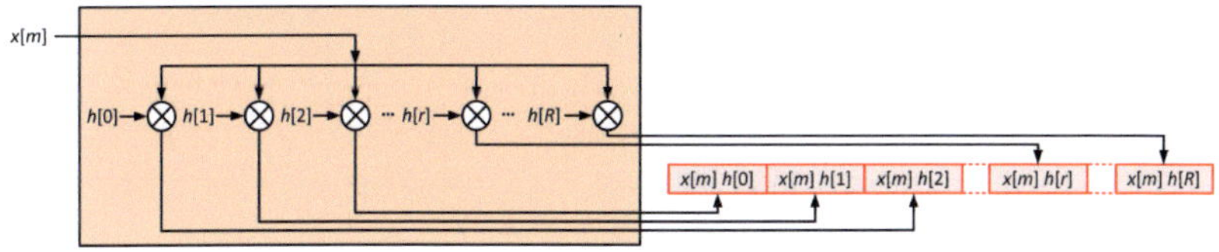

Fig. 4.16 Impulse response generator. A copy of the input sample is multiplied by each of the filter coefficients $h[r]$ representing the sampled IR of the filter with order R and $R + 1$ filter coefficients. The multiplication with $h[r]$ leads to a finite number of products $x[m]$ $h[r]$ that can be stored in look-up tables (LUT) thereby avoiding resource-hungry multiplications.

With N copies of the IR generator of Fig. 4.16 we process N samples in parallel and hence reduce the clock speed of the DSP by a factor N. Processing of $N = 4$ samples per clock cycle (cycles are separated by dashed lines) is shown in Fig. 4.17. The clock waveform (purple) is depicted vertically on the left-most side of Fig. 4.17. Processing is performed at the rising edges of the clock. In each clock cycle $N = 4$ input samples $x[m]$ are fed to the IR generators of Fig. 4.16 leading to a block of $N = 4$ sampled and weighed IRs $x[m]\,h[r]$ (as depicted in the middle of Fig. 4.17). In this example, the IR of the filter is represented by $R + 1 = 8$ filter coefficients or filter taps (filter order $R = 7$). In order to prepare the computation of the filter output $y[m]$, the IRs are arranged in a two-dimensional (2D) array (right). After flipping the weighed IRs upside down, each IR within a row (elements $[r]$) is rotated counter-clockwise and mapped to a column (elements $[m]$) of the 2D array such that adjacent responses $x[m]\,h[r]$ are shifted by one sample with respect to each other. The output $y[m]$ is then obtained by summing all samples that are located in the same row of the 2D array.

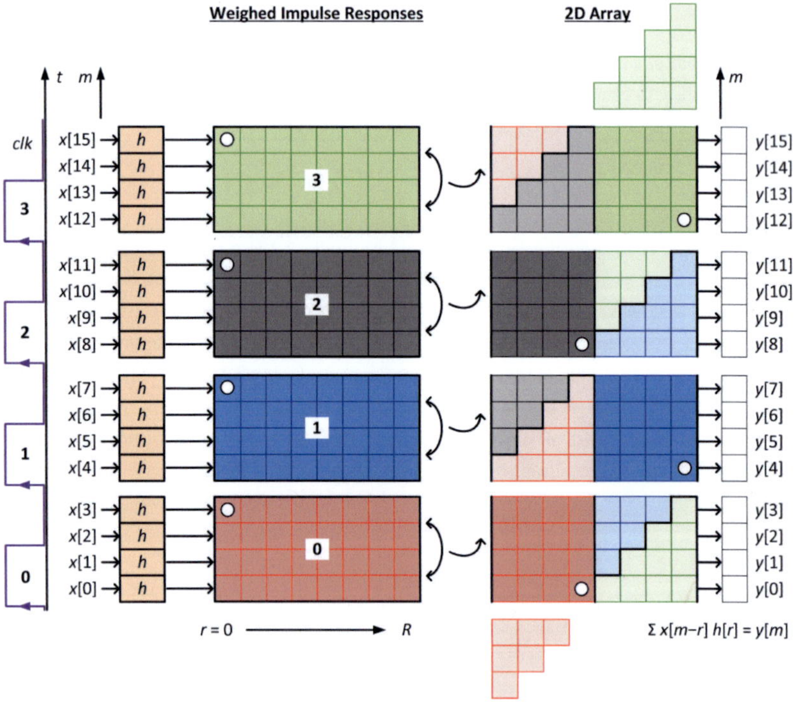

Fig. 4.17 Parallel FIR filter design where $N = 4$ input samples $x[m]$ are processed within each clock cycle (*clk*, left vertical time axis t) to produce four output samples $y[m]$ with the summation $y[m] = \sum x[m - r]\,h[r]$, Eq. (2.5.13). Each input sample is multiplied with the filter impulse response (IR) $h[r]$ using the IR generators described in Fig. 4.16. In this example, the filter response is represented by $R + 1 = 8$ filter coefficients. In order to prepare the computation of the filter output $y[m]$, the pulses are arranged within a two-dimensional (2D) array (right, flipped upside down and rotated counter-clockwise by 90°) with each IR at time m being delayed by one sample with respect to the previous IR at $m - 1$. The output $y[m]$ is then obtained by summing all samples that are located in the same row of the 2D array.

From Fig. 4.17 it can be seen that some samples of the weighed IRs spread into adjacent processing cycles. Buffer areas in the 2D array (lighter colors) are used to correctly account for these samples. To compute the $N = 4$ output samples of clock cycle 2 (black), input samples from the blue (cycle 1), black (cycle 2) and green (cycle 3) colored IRs need to be available. Therefore output values $y[8]...y[11]$ have to be computed within the following clock-cycle 3 leading to a latency of one cycle. In Fig. 4.17, it can be further seen that some samples of the red (cycle 0) and the green IRs (cycle 3) extend beyond the boundaries of the 2D array. These samples fit to the opposite sides of the array (rows $m = 13...15$ for the red samples and rows $m = 0...3$ for the green samples). With each additional clock cycle (4, 5, ...) the samples in the physical storage areas 0, 1, 2, 3 are replaced by the following scheme: cycle 4→area 0, cycle 5→area 1, cycle 6→ area 2, cycle 7→ area 3, cycle 8→ area 0, ... Four storage areas were chosen in Fig. 4.17 for simplifying the explanation. However, since only samples from three cycles are required for computing $y[m]$, the 2D array needs containing only three physical storage areas 0, 1, 2 with $R + 1 = 8$ columns and 12 rows.

Besides processing multi-gigabit data streams in real-time, the described parallel FIR filter design allows changing the symbol rate and the bandwidth by varying the oversampling factor q during runtime. To this end, only the filter coefficients $h[r]$ in the IR generators of Fig. 4.16 (brownish boxes h in Fig. 4.17) need to be changed. The granularity with which q can be changed scales with the number N of samples being processed in parallel. This is similar to the procedure for OFDM, where q is changed by nulling more or less of the outer SCs, and the granularity therefore scales with the number of SCs. To illustrate the flexible adaption of symbol rate and bandwidth through a change of q, we depict a single clock cycle 0 and its associated storage area 0 from Fig. 4.17 for $N = 4$ and $q = 1$, 2, 4, and 4 / 3, see Fig. 4.18. The IR generators h are now subscripted with q for $q = 1$, 2, 4, and with A, B, C for $q = 4 / 3$. The situation in Fig. 4.17 with an oversampling factor $q = 1$ is reproduced in Fig. 4.18(a). For an oversampling factor $q = 2$, Fig. 4.18(b), the filter coefficients $h[r]$ of every other IR generator (Fig. 4.16) are set to zero. As a consequence, only every second input sample $x[m]$ transmits data, so that the data rate and the bandwidth of the transmitted signals is reduced by a factor of $q = 2$. In analogy to $q = 2$, an oversampling factor $q = 4$ is obtained by setting the filter coefficients $h[r]$ in three out of the four IR generators to zero, see Fig. 4.18(c). Finally, a rational oversampling factor $q = 4 / 3$ can be realized, see Fig. 4.18(d). This time the coefficients $h[r]$ in the IR generators are adjusted according to Fig. 4.14(a), where one distinguishes between three identical but differently sampled impulse responses (black, red, and blue). The exchange of filter coefficients can be realized in a single clock cycle. It should be noted, however, that a change of q affects the duration $T = (R / q)\, T_s$ of the impulse response which is represented by a fixed number of $R + 1$ coefficients. For q close to 1, any infinitely extended impulse response is much better approximated than for a q close to N.

This affects the steepness of the signal band slopes and thus the potential SE in a way similar to OFDM when nulling outer SCs to effectively increase q.

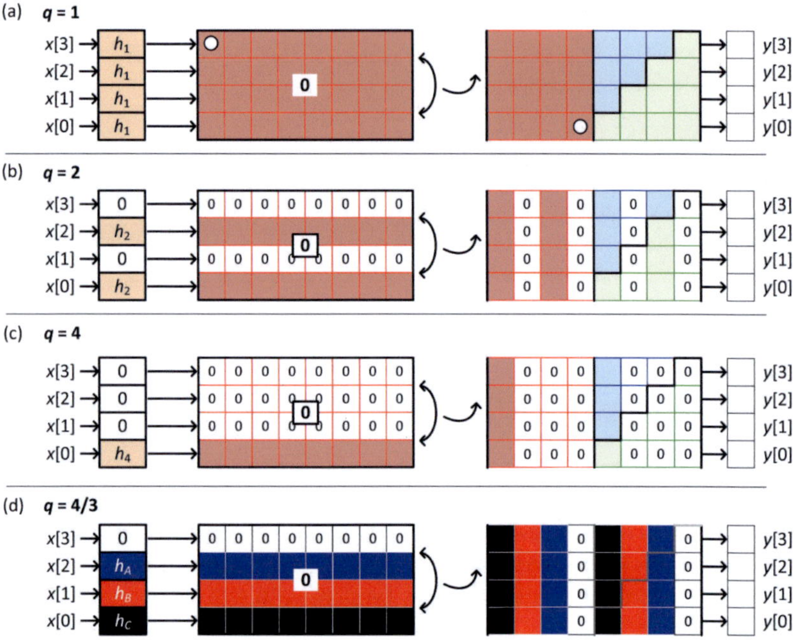

Fig. 4.18 Flexible adaptation of the symbol rate and bandwidth through changing the oversampling factor q while keeping the sampling rate f_s constant. For clarity, only clock cycle 0 according to Fig. 4.17 is shown. (**a**) Clock cycle 0 of Fig. 4.17 is reproduced, $q = 1$ (**b**) For an oversampling factor $q = 2$ every other IR generator must produce zeros as impulse response, thereby reducing the data rate by a factor of 2. (**c**) To generate signals with $q = 4$, three of the 4 pulses processed in parallel are set to zero. (**d**) A rational oversampling factor such as $q = 4/3$ can be realized by adjusting the IR generator coefficients $h_{A,B,C}$ according to Fig. 4.14(a).

In order to prove the adaptability of our signal processing with respect to symbol rate and signal bandwidth we use a filter with order $R = 64$ and simulate the VHDL-description (very high speed integrated circuit hardware description language) on the register transfer level (RTL), a design abstraction which models the signal flow of synchronous digital circuits. While the filter coefficients are varied to cover the three practical scenarios indicated in Fig. 4.18(b)–18(d), the clock rate remains constant. From the generated signals we derive the ensemble-averaged power spectra using a fast Fourier transform (FFT). Fig. 4.19 shows the averaged power spectra as a function of frequency normalized with respect to the sampling frequency f_s. As expected, the spectrum for $q = 4$ is confined between $-0.125 f_s$ and $0.125 f_s$, i.e., the bandwidth is reduced to $f_s / 4$, see Fig. 4.19(a).

Changing the filter coefficients for an oversampling factor $q = 2$ increases the symbol rate and the bandwidth to $f_s / 2$, see Fig. 4.19(b). Finally, we adapt the filter coefficients for an oversampling of $q = 4 / 3$ and obtain a spectrum confined between $-0.375 f_s$ and $0.375 f_s$ and thus a signal bandwidth of $f_s / (4 / 3)$, Fig. 4.19(c).

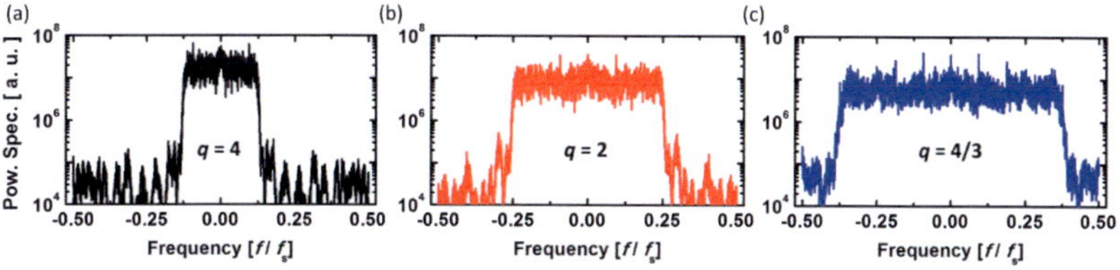

Fig. 4.19 Ensemble-averaged power spectra obtained from VHDL simulations as a function of frequency normalized to the sampling rate f_s. All Nyquist signals are generated with the same FIR filter structure of order $R = 64$. (**a**) Spectrum for a signal oversampling of $q = 4$. The signal bandwidth is $f_s / 4$. (**b**) Spectrum for an oversampling $q = 2$. The signal bandwidth is $f_s / 2$. (**c**) Spectrum for an oversampling $q = 4 / 3$. The signal bandwidth is $f_s / (4 / 3)$.

Resource requirements: The previously described real-time parallel processing has been implemented in VHDL, synthesized, and then evaluated in terms of resources. We implemented FIR filters of different orders R and with an oversampling factor $q = 4 / 3$. The corresponding impulse response duration T for a filter of order R is $T = (R / q)\, T_s$. The higher the order R, the better a sinc-impulse response is approximated [79]. All evaluated filter designs are optimized with respect to minimum resource requirements and maximum speed, i.e., no flexible adaptation of q is supported at this point. Since we used the filters for an M-ary QAM Tx, we realized the IR generators of Fig. 4.16 with LUTs for avoiding multiplications. For the final FPGA design a number of $N = 128$ samples are processed in parallel. Table 4.1 shows the resource utilization of the FPGA for a Nyquist pulse-shaper of order R that can flexibly switch between QPSK and 16QAM. We depict the most significant resources, i.e., the number of slice registers and slice LUTs. To give a quantitative estimate of what can be achieved with state-of-the-art FPGAs, we relate the resources to one of the biggest available Virtex 7 FPGAs (XC7VX980T). In this work, however, we were restricted by the number of slice registers and slice LUTs provided by two Virtex 5 FPGAs (XC5VFX200T). Therefore we used a filter of order $R = 32$ for the experiments discussed later on, since it can be handled by the Virtex 5 FPGA while still meeting the timing constraints with on-chip clock frequencies of up to $f_s / N = 28\text{GHz} / 128 = 218.75$ MHz.

We now compare the resource requirements of processing with the standard two-fold oversampling [79][129] q_2 to processing using (4/3)-fold oversampling $q_{4/3}$. For a filter order $R = 128$ and a degree of parallelization $N = 128$ we find that the number of slice registers increases by a factor of 2.8 and the number of slice LUTs increases by a factor of 2.7. However, the number of symbols processed within each clock cycle also increases by a factor of $q_2 / q_{4/3} = 3 / 2$, and so does the duration T of the impulse response. With these numbers we calculate the amount of slice registers and slice LUTs required to process a single symbol for a fixed impulse response duration T. We find that a 1.5-fold (50 %) increase in speed comes at the moderate price of a 14 % (slice registers) and 17 % (slice LUTs) increase of resources.

Table 4.1 FPGA resource utilization for FIR filters of different orders R and impulse response duration T.

Resource	R = **16** (T = 12 T_s)	R = **32** (T = 24 T_s)	R = **64** (T = 48 T_s)	R = **128** (T = 96 T_s)
Slice registers (% of XC6 / 2×XC5)	29300 (4 % / 12 %)	60120 (8 % / 24 %)	109940 (14 % / 45 %)	210320 (28 % / 86 %)
Slice LUT (% of XC6 / 2×XC5)	25180 (6 % / 10 %)	50360 (14 % / 20 %)	93120 (26 % / 38 %)	175080 (48 % / 71 %)

4.2.2.5 Experimental Setup

The real-time Nyquist pulse Tx, Fig. 4.20, comprises two synchronized Xilinx XC5VFX200T FPGAs, two Micram 6 bit DACs, optional 12.3 GHz anti-alias low-pass filters, and an optical I/Q-modulator. The real-time generated Nyquist sinc-pulses are encoded on an external cavity laser (ECL) operated at a center wavelength of 1550 nm.

For experiments with transmission (switch position 1), the link is formed by an erbium doped fiber amplifier (EDFA 1) together with a polarization emulation setup where the signal is split, delayed by 5.4 ns with respect to each other, and combined in orthogonal polarizations. Measurements to evaluate the performance of the dynamic computational precision are performed with a transmission link of up to four 75 km spans of SSMF (attenuation: 0.19 dB / km, dispersion: 16.8 ps /nm /km) with in-line EDFA amplification. For the investigation of PDM-64QAM oversampled by $q = 4 / 3$, the transmission link comprises 100 km ULAF (attenuation: 0.18 dB / km, dispersion: 20.2 ps /nm /km). The input power to the Rx EDFA 2 is adjusted to sweep the optical signal-to-noise ratio (OSNR) (switch position 2), measured with an optical spectrum analyzer (OSA) (reference bandwidth: 0.1 nm). A 0.6 nm filter removes out-of-band EDFA noise. For operating the Agilent optical modulation analyzer (OMA) optimally, EDFA 3 is employed (switch position 2). Finally, the OMA coherently receives the signals using a second, inbuilt ECL and real-time oscilloscopes. Further processing including error vector magnitude (EVM) and bit error ratio (BER) computation is performed offline.

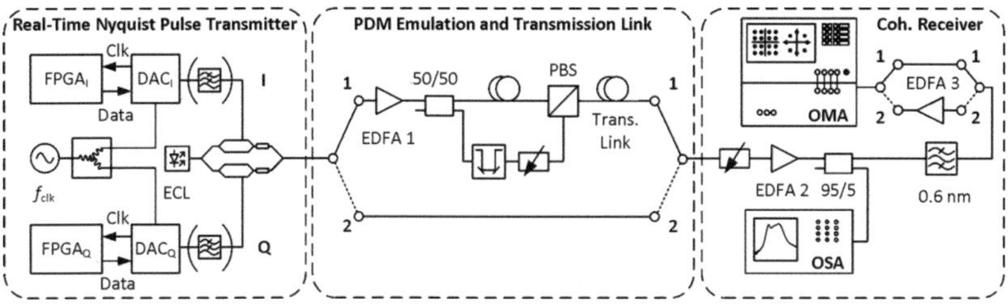

Fig. 4.20 Experimental setup for real-time Nyquist sinc-pulse shaping. The Tx comprises two synchronized FPGAs, two DACs, optional image rejection filters, and an optical I/Q-modulator. Data are encoded on an ECL at 1550 nm. For transmission experiments (switch position 1): EDFA 1 and PDM emulation followed by either up to four 75 km spans of SSMF and EDFAs (for the evaluation of the dynamic computational precision) or by 100 km ULAF (for evaluation of $q = 4 / 3$ oversampled PDM-64QAM transmission). Signals are recovered with a coherent receiver (EDFA 2 and Agilent OMA). For QPSK and 16QAM measurements with oversampling factor $q = 4 / 3$ only, switch position 2 bypasses the link. The OSNR is adjusted by varying the input power to EDFA 2. In this case, we keep the OMA at optimum input power with a gain-controlled EDFA 3.

4.2.2.6 Experimental results

In the following we present the results obtained with the setup of Fig. 4.20. First, we investigate the performance of dynamically adjusted precision for PDM-16QAM and PDM-64QAM signals transmitted over up to 300 km of SSMF. Second, Nyquist signals are generated with an oversampling factor $q = 4 / 3$. We vary the sampling rate of the DACs resulting in QPSK and 16QAM Nyquist signals with symbol rates of 15 GBd, 18.75 GBd, and 21 GBd. We further demonstrate PDM-64QAM at 21 GBd and transmit the signal over 100 km ULAF. For all measurements we record the BER counting at least 100 errors per measurement and/or the EVM derived from >50,000 received symbols.

Dynamic computational precision: Finite duration impulse response filters with conventional, "standard" DSP (fixed effective word length) and with dynamically adjusted DSP word length (with variants ÷ and ×) are implemented on the register transfer level (RTL) using VHDL. All filters in this section are of order $R = 64$ and process two-fold oversampled signals ($q = 2$). The overall chip (FPGA) utilization for the filters with different DSP precision is depicted in Table 4.2.

Table 4.2 FPGA resource utilization for FIR filters with and without dynamically adjusted DSP word length.

Resource	Fixed word length (WL)	Dynamically adjusted WL ÷-variant	Dynamically adjusted WL ×-variant
Slice registers (% of XCV5FX200T)	39,040 (32 %)	39,040 (32 %)	36,600 (30 %)
Slice LUT (% of XCV5FX200T)	31,250 (25 %)	32,500 (26 %)	30,000 (24 %)

Subsequently, the filters are employed in our real-time Nyquist Tx. Measured formats include 14 GBd PDM-16QAM and 12.5 GBd PDM-64QAM. Nyquist pulses are transmitted over up to four spans of 75 km SSMF. A measured (black) and VHDL-simulated (white) spectrum of a 16QAM signal is depicted in Fig. 4.21(a). The signal bandwidth is close to 14 GHz for a symbol rate of 14 GBd. The observed noise floor is due to quantization noise of the 6 bit Tx DACs and the 8 bit Rx ADC (OMA oscilloscope). We further measure the BER for PDM-16QAM and PDM-64QAM signals, for all filter implementations and for different fiber spans, Fig. 4.21(b). It can be concluded from Fig. 4.21 and Table 4.2 that signals generated with dynamically adjusted word length from 6 bit to 11 bit give superior signal quality without increasing the processing complexity. As expected, the (×)-filter shows best performance requiring similar resources on the FPGAs as the standard fixed precision DSP.

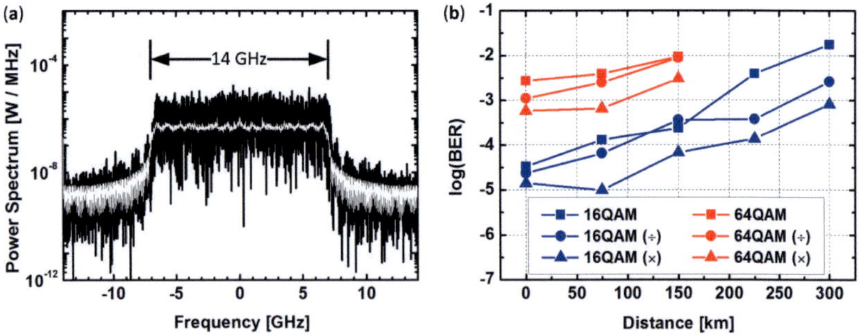

Fig. 4.21 Experimental results for Nyquist shaped PDM-16QAM (14 GBd) and PDM-64QAM (12.5 GBd) transmitted over various spans of SSMF. (**a**) Measured (black) and simulated (white) spectra for 16QAM signals. The spectral noise floor is due to quantization noise, and can be removed by analog low-pass filters. (**b**) Measured BER of both modulation formats and fiber spans for standard and dynamically adjusted DSP word lengths as described in Fig. 4.13(c). As expected, the (×)-design shows best performance.

Rational oversampling: We measure the spectra, the BER, and we estimate the BER from the measured EVM [81][82] for Nyquist sinc-pulses generated with a real-time pulse-shaper (R = 32) and with an oversampling factor $q = 4/3$ for various OSNR, Fig. 4.22. Results obtained with (without) image rejection filters are colored black (red). The measured signal spectra in Fig. 4.22(a) show an out-of-band suppression of 30 dB and a bandwidth close to 21 GHz for 21 GBd signals. The image spectra Fig. 4.22(a) (red) are effectively removed by the filters (black). Fig. 4.22(b)–22(d) depicts the measured BER (squares) and BER as calculated from measured EVM (lines).

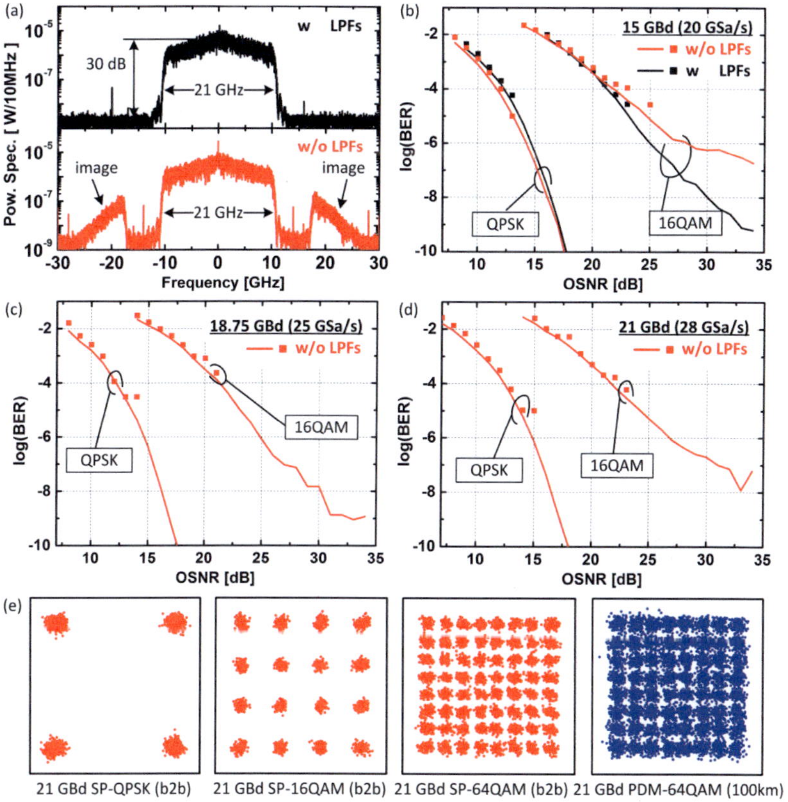

Fig. 4.22 Experimental results for real-time Nyquist pulse shaping with filter order $R = 32$. **(a)** 21 GBd 16QAM spectra measured with low pass filters (LPF) used for image rejection (black) and without (red). **(b)** – **(d)** BER (squares) and BER estimated from measured EVM (lines) as a function of OSNR for QPSK and 16QAM with (black) and without (red) image rejection filters. **(e)** Constellation diagrams obtained back-to-back for a single polarization (SP, red) and for PDM-64QAM after transmission over 100 km (blue).

For symbol rates beyond 15 GBd the available image rejection filters introduce a penalty due to the non-optimum cut-off frequency of 12.3 GHz. Due their suboptimal frequency response [130] for Nyquist signals with these symbol rates, measurements in Fig. 4.22(c) and 22(d) are taken solely without filters. Recently, it has been shown that carefully designed low-pass filters (LPF) allow for oversampling factors as low as $q = 1.15$ [131]. Fig. 4.22(e) shows constellation diagrams for 21 GBd single polarization (SP) signals (red) at highest possible OSNR. For SP-64QAM (red) and in a back-to-back configuration, a BER = 1.84×10^{-3} is achieved. After 100 km transmission of PDM-64QAM (blue) at a bit rate of 252 Gbit/s the resulting raw BER of 1.43×10^{-2} can still be improved by state-of-the-art forward error correction (FEC) [132].

4.3 Real-Time OFDM Transmitters

Now real-time OFDM signal generation is discussed. Again state-of-the-art FPGAs are used to process OFDM signals with data rates beyond 100 Gbit/s. First a LUT realization of an IDFT is discussed and then an IFFT is employed.

4.3.1 Real-time OFDM transmitter beyond 100 Gbit/s

Real-time OFDM transmitter beyond 100 Gbit/s

R. Schmogrow, M. Winter, D. Hillerkuss, B. Nebendahl, S. Ben-Ezra, J. Meyer, M. Dreschmann, M. Huebner, J. Becker, C. Koos, W. Freude, and J. Leuthold

Opt. Express 19(13), pp. 12740–12749 (2011). [J15]

4.3.1.1 Introduction

Orthogonal frequency division multiplexing (OFDM) emerged as a candidate for high performance optical communications in both long-haul and access network scenarios [23][133]. First experiments have been shown using offline processing for transmitting and receiving OFDM signals over optical fiber [117][134]. However, practical applications call for transmitters and receivers that perform real-time data processing. Application specific integrated circuits (ASIC) or field programmable gate arrays (FPGA) together with high-speed digital-to-analog (DAC) and analog-to-digital converters (ADC) enable the implementation of high-speed real-time OFDM transmission. While ASICs are compact and efficient, they require extensive development time and budget. Conversely, FPGAs are ideal for fast prototyping. Despite the high complexity of real-time OFDM, experiments demonstrating real-time OFDM transmitters [34][135][32][37] and receivers [36][35] have recently been shown.

In this paper we introduce the first single polarization 101.5 Gbit/s real-time OFDM transmitter modulating 58 subcarriers with 16QAM. Contrary to other implementations that rely on the fast Fourier transform (FFT) algorithm by Cooley and Tuckey, we concentrated on a 64-point inverse discrete Fourier transform (IDFT), which we implemented in a highly efficient form on an FPGA. The paper is an in-depth report of the first real-time 100 Gbit/s OFDM transmitter demonstration recently presented at OFC [135].

4.3.1.2 Experimental Setup

The real-time OFDM transmitter (Tx) comprises two Xilinx Virtex 5 FPGAs, two high-speed Micram DACs with 6 bit resolution, and an optical IQ-modulator. We modulate a continuous wave (CW) external cavity laser (ECL) with in-phase (I) and quadrature-phase (Q) data as shown in Fig. 4.23. A 28 GHz sinusoidal clock signal is split, phase aligned and fed to the DACs which provide the reference clock signal for the FPGAs. Within the FPGAs, a complex OFDM signal is calculated from a pseudo random bit sequence (PRBS, $2^{15} - 1$) in real-time and passed to the DACs. The DAC outputs deliver 28 GSa/s and feed the IQ-modulator for generating a 101.5 Gbit/s single polarization coherent optical OFDM signal.

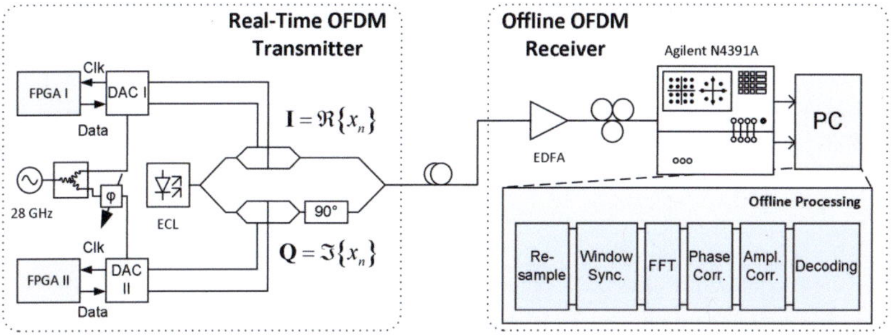

Fig. 4.23 Experimental setup. Real-time OFDM transmitter comprising FPGAs, DACs, and an external cavity laser (ECL) source with an optical IQ-modulator. Both FPGA boards are fed with identical spectral data sequences X_k (in fact, these pseudo-random data are generated on the FPGA boards themselves). The boards generate the corresponding complex OFDM symbols xn by an inverse Fourier transform, Eq. (2.5.14). FPGA1 and FPGA2 then feed the real part of x_n to DAC1 for the I-channel and the imaginary part of xn to DAC2 for the Q-channel modulation, respectively. A short standard single-mode fiber connects the transmitter to an offline OFDM receiver comprising an EDFA, a polarization controller and an Agilent OMA, where the signal is sampled at 80 GSa/s. The received signal is then processed offline using standard OFDM algorithms.

The receiver (Rx) comprises an erbium doped fiber amplifier (EDFA) that boosts the optical OFDM signal in order to set the power to the receiver's optimum operating point. The signal is then received by an Agilent modulation analyzer (OMA) and sampled by a 20 GHz real-time oscilloscope with 80 GSa/s on two channels simultaneously. The received data are processed offline using standard OFDM receiver algorithms. After a fast Fourier transform (FFT), phase drifts and time-linear phase variations are compensated through the phase information provided by the pilot tones. For real-time processing, additional training sequences would facilitate channel estimation and equalization. Finally, we equalize the frequency-dependent amplitude of the subcarriers (SC), decode the SC signals, and evaluate the error-vector magnitude (EVM).

4.3.1.3 Digital Signal Processing

One of the main challenges in generating single-carrier quadrature amplitude modulated (QAM) signals and OFDM waveforms is due to demanding digital signal processing (DSP) requirements that come with any real-time OFDM transmitter. Basic DSP schemes can be found in several publications [136][33], but the actual implementation and optimization of the DSP blocks is crucial for high performance transmitters.

General processing within FPGA: In order to generate a complex OFDM waveform, we use two FPGAs to calculate the real and imaginary part of the OFDM signal x_n, respectively. Each FPGA generates pseudo-random binary sequences (PRBS) for randomly generating complex spectral data X_k, which then form the OFDM symbol via an inverse Fourier transform, Eq. (2.5.14) and Fig. 2. The PRBS generators on both FPGAs are synchronized, i.e., they generate the same random complex data X_k in synchronism. For M-QAM, a number of $\log_2 M$ consecutive PRBS bits determine X_k within one OFDM symbol for subcarrier k.

The FPGA-internal DSP blocks are shown in Fig. 4.24 including the interface to the high-speed DAC. Binary data are fed to the inverse discrete Fourier transform (IDFT) block which is later on described in more detail. After the IDFT block, a clipping and rescaling module trims the IDFT output to the 6 bit physical resolution of the DAC. For our investigations we choose the 16QAM format (4 bit) and a number of 64 subcarriers. The relative amount of signal clipping and scaling is selected such that clipping errors and quantization noise have minimum impact on the output signal [81][82]. To this end, we employ a 64-point IFFT with a numerical precision of 53 bit (double precision floating point, 64 bit word length), which generates an ―ideal‖ random sequence of OFDM symbols. For various combinations of clipping levels and signal scaling we calculate the minimum error vector magnitude (EVM) for a target DAC resolution of 6 bit. Under these conditions, the optimum EVM amounts to 2.7% and the probability is as large as 93% that the compound OFDM signal's values are located inside the DAC quantization window.

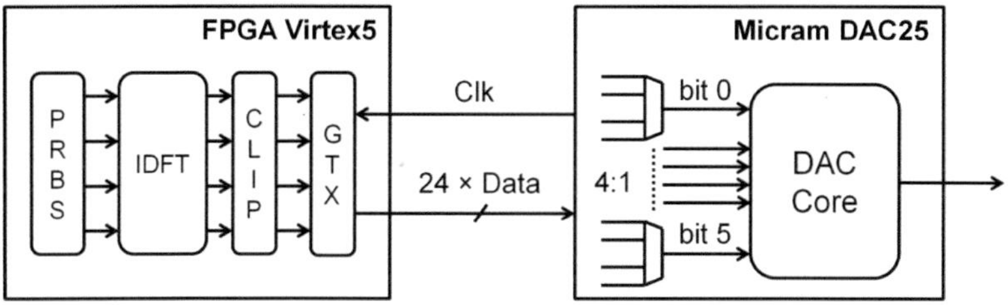

Fig. 4.24 FPGA and high speed DAC with digital signal processing (DSP) blocks and GTX interface. A pseudo random binary sequence (PRBS) generator feeds the input of the IDFT core which produces the representation of the time domain OFDM signal. A clipping and rescaling module (CLIP) trims the signal to the physical resolution of the Micram DAC. The FPGA's 24 high-speed synchronized GTX transmitters drive the DAC at 7 Gbit/s each. Onboard 4:1 multiplexers before the DAC core translate to an output sampling rate of 28 GSa/s with 6 bit of physical resolution.

We used Xilinx Virtex 5 FPGAs for the experiments. They provide 24 high-speed transmitters (referred to as GTX) that drive the inputs of the Micram DAC at 7 Gbit/s on each line. Multiplexers enhance the output sampling rate by a factor of 4 resulting in a sampling rate of 28 GSa/s at the DAC's output. The DAC provides a reference clock to the FPGA which is used for the DSP blocks. Since the FPGA gains its speed through parallelization, a relatively low clock rate suffices to generate the data. Therefore, the full-rate DAC sampling clock of 28 GHz is divided by a factor of 128 resulting in an FPGA reference clock rate of 218.75 MHz.

Inverse discrete Fourier transform: The key element of an OFDM transmitter is the transformation of frequency domain data into a time domain waveform. As most real-time OFDM transmitters exploit the strength of the IFFT algorithm, we follow a different approach by directly implementing the inverse discrete Fourier transform (IDFT)

$$x_n = \sum_{k=0}^{N-1} X_k \, \mathrm{e}^{\mathrm{j}2\pi \frac{kn}{N}}, \quad k,n = 0,1,\ldots,N-1, \quad M \text{ complex data symbols } X_k \text{ per } k, \tag{2.5.14}$$

in a highly efficient manner. Here, x_n represents the nth time domain sample, and X_k is the set of M complex modulation coefficients (data symbols) available for the k th complex harmonic subcarrier. For a time domain sample x_n, the sum over all N subcarriers has to be calculated. A schematic of the on-chip design is depicted in Fig. 4.25. Since digitally modulated subcarriers are represented by a discrete set of M coefficients X_k only, all possible variations of modulated subcarriers $X_k \exp(\mathrm{j}2\pi kn/N)$ can be stored using look-up tables (LUT), thereby avoiding complex multiplications at runtime. For simplicity, Fig. 4.25 shows a selection of $4 \times M$ subcarrier waveform LUTs out of $N \times M$ in a waterfall display. Each LUT stores N samples for M possible waveforms of a fixed subcarrier $k = 0, 1, \ldots, N-1$. The figure displays the LUTs for the real part of a modulated SC only. The calculation of the imaginary part follows an analog scheme. This arrangement lends itself to parallelization employing two synchronized FPGAs. The OFDM output signal is obtained by summing all N subcarrier samples for each position n.

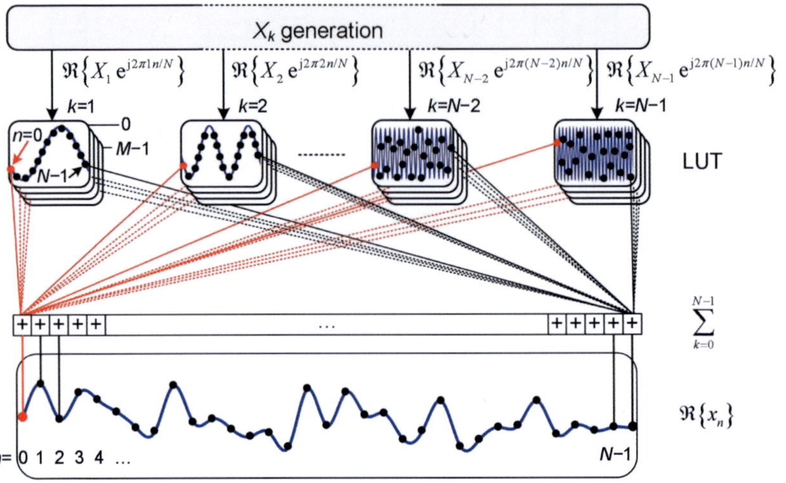

Fig. 4.25 The FPGA's internal realization of the IDFT operation. All M possible products of modulation coefficients X_k and k^{th} complex harmonic are sampled, quantized, and stored within look-up tables (LUT). There are LUTs for the real part and for the imaginary part of each complex harmonic. For simplicity we show only one set of LUT, namely the one for the real part which corresponds to the inphase component I. An onboard PRBS generator selects symbols to be transmitted by addressing the corresponding LUTs. For each sample of each of the subcarriers, an addition forms the output samples (k = 0…N–1) of the transformed signal. To improve the computation efficiency additions are realized by a binary adder tree with $\log_2 N$ stages.

Although this implementation avoids real-time complex multiplications, the extensive use of LUTs and adders could seem inefficient at first sight. However, a closer inspection of Eq. (2.5.14) reveals redundancies. This is discussed in the following with reference to Fig. 4.26.

First, we exploit the periodicity of the N modulated subcarriers $X_k \exp(j2\pi kn/N)$ for a given k with respect to time-sample index n. This periodicity is discussed with the help of Fig. 4.26(a) where all $N = 64$ $X_k \exp(j2\pi kn/N)$ coefficients for $k = 6$ and $X_k = 1$ are plotted. It can be seen that the solid line interpolating on the sample dots has a subcarrier period $p_k = N / k$, which need not necessarily be an integer. Further, the samples repeat with the periodicity p_n. To extract this periodicity, we need to find the greatest common divisor (GCD) of N and k. If N can be factored into two integer numbers such that $N = k_3\, k_2$, and k can be factored into $k = k_1\, k_2$ with $k_1, k_2, k_3 = 1, 2, \ldots, N$, then the maximum number $k_2 > 1$ is the (non-trivial) GCD of N and k, GDC(N, k) = k_2, and the quantities N and k are called (non-trivially) commensurable with respect to k_2. We find that this period is $p_n = N / k_2$. Hence, for each possible data symbol X_k, only the k_2^{th} part of the maximum number of N samples needs to be processed separately. If the GCD of N and k is k itself, see Fig. 4.26(b), then $p_n = p_k$, and only p_k values out of N are different. If $k = 0$, then only one sample has to be processed.

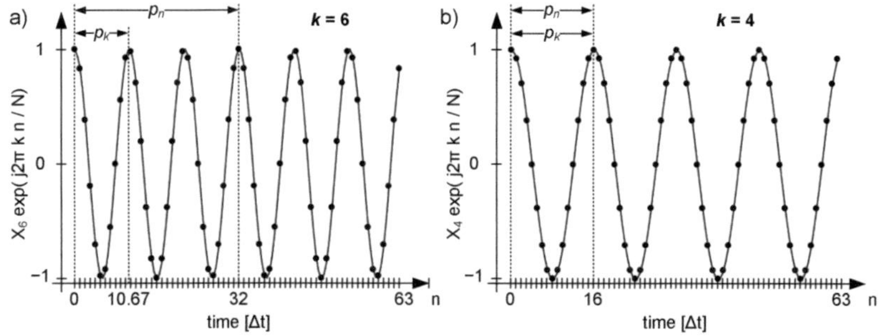

Fig. 4.26 Real output sample values of a 64-point IDFT for subcarriers (a) $k = 6$ and (b) $k = 4$ with the specific modulation symbols $X_6 = X_4 = 1$. The time sample number n on the horizontal axis can be interpreted in units of an arbitrary time interval Δt. The solid blue line represents the interpolated —physical subcarrier time function, if n is assumed to be continuous. (a) Subcarrier $k = 6$ is doubly periodic within the IDFT window. It has the —physical non-integer period $p_k = 64 / 6 = 32 / 3 \approx 10.67$ and a minimum integer period $p_n = 64 / 2 = 32 = 3 \times p_k$. Therefore only 32 samples have to be processed. (b) Subcarrier $k = 4$ is singly periodic with a minimum integer period $p_k = 64 / 4 = 16 = p_n$, so that only 16 samples out of 64 need to be processed.

If redundancies are disregarded, the total number P of samples to be processed with the help of the LUT is, according to Eq. (2.5.14),

$$P = N^2 \tag{2.5.15}$$

If sample periodicities are exploited, then the total number of samples to be processed can be reduced to

$$P = 1 + \sum_{k=1}^{N-1} \frac{N}{\mathrm{GCD}(N,k)} \tag{2.5.16}$$

For the case where $N = 2^q$ is a power q of 2, maximum savings of 1 / 3 can be achieved for $N >> 1$. This may be derived from Eq. (2.5.16) or is directly proved as follows: For half of the subcarriers, i.e., for ½ 2^q subcarriers, the carrier index $k = k_o$ is odd, so N and k_o have the GCD $k_2 = 1$. All 2^q samples have to be processed, because no repetition of the basic interval p_k can be found. For the even-numbered subcarriers, one half of them, namely ½ 2^{q-1} subcarriers, have the GCD $k_2 = 2$. Therefore, only the k_2^{th} part of the 2^q samples is to be processed, i.e., 2^{q-1} samples. The other half of the even-numbered subcarriers is again split in two. One of these groups, namely ½ 2^{q-2} subcarriers, has a GCD of $k_2 = 4$ with respect to N and k, therefore 2^{q-2} different samples have to be processed. If we proceed with splitting into subcarrier groups, we end up with a sequence of 2^{q-0}, 2^{q-1}, 2^{q-2}, …, 2^{q-q} samples which need processing. We find the total number of samples to be processed per group by multiplying with the number of subcarriers in each group, namely ½ 2^{q-0}, ½ 2^{q-1}, ½ 2^{q-2}, …, ½ $2^{q-(q-1)}$, ½ 2^{q-q}. In this sequence, the last term deviates from the rest, because in this subcarrier group (as in the last but one group), one subcarrier has to be counted. The total number P_q of samples to be processed for the case where $N = 2^q$ is then

$$P_q = 1 + \sum_{m=0}^{q-1} 2^{q-m} \times \tfrac{1}{2} 2^{q-m} = 1 + \tfrac{1}{2} 2^{2q} \sum_{m=0}^{q-1} 4^{-m} = 1 + \tfrac{1}{2} N^2 \sum_{m=0}^{q-1} 4^{-m}, \ \lim_{q\to\infty} P_q = \tfrac{1}{2} N^2 \frac{1}{1-\frac{1}{4}} = \tfrac{2}{3} N^2 . \quad (2.5.17)$$

For the limiting case that $N = 2^q$ is large, the number of samples to be processed is well approximated by $P_{q>>1} \approx 2\,N^2 / 3$. Therefore, a maximum of 33 % complex adders and LUTs can be saved without compromising computation accuracy.

Second, only data transmitting subcarriers have to be processed using our described IDFT whereas the IFFT must process all N subcarriers at all times. Thus additional savings in complexity are obtained when inserting frequency guard bands as is regularly done in OFDM systems. For instance, we use four pilot tones that have been assigned to the odd subcarriers $k = 7, 21, 43, 57$ which correspond to the positions 7, 21, −21, −7 defined in the IEEE 802.11-2007 WLAN standard. Instead of processing $4 \times M \times N$ samples for the pilot tones, only 64 samples suffice to represent the full information.

Third, even more LUT storage space can be saved by taking advantage of symmetry relations between the M modulation coefficients X_k. Given that digital modulation formats lead to (in general complex) modulation coefficients X_k with only a discrete number of different phases (e.g., 4 for QPSK), another 75% of LUT storage space can be saved resulting in an overall reduction of 83.5%. The principle of this technique is shown in Fig. 4.27 for a multilevel phase modulation of the first subcarrier (SC 1). For a better understanding, Fig. 4.27(b) explicitly shows the periodic continuation of the basic modulated subcarrier. Instead of storing all M waveforms corresponding to the M symbols as shown in Fig. 4.27(a), only $M / 4$ waveforms have to be stored since only they consist of different values. A pointer selects an offset within the LUT corresponding to a phase shift of the waveform (e.g., $\varphi = \pi / 4$).

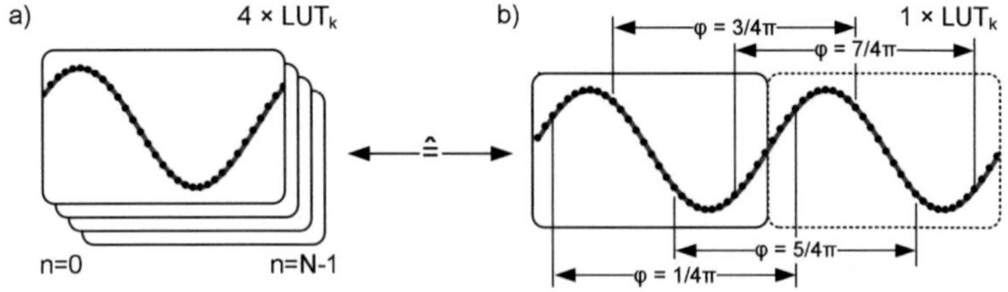

Fig. 4.27 LUT contents for subcarrier k and a set of $M = 4$ symbols (QPSK). (a) LUT with N values for each symbol in the constellation. (b) Only one LUT needs to be stored. Phases exceeding the stored range are folded back due to the function's periodicity.

Comparisons with other algorithms such as the fast Fourier transform heavily depend on the specific implementation on the FPGA and the hardware platform. Therefore a general estimation of performance and utilized chip area cannot be easily given. This would require implementing competing algorithms and evaluating maximum operation speed, calculation accuracy, and chip area utilization. For higher order IDFTs LUT space becomes an issue. Nevertheless, a 128-point IDFT following the described procedure is not out of reach. Naturally, the storage space doubles. However, in the current design two 64-point IDFTs run in parallel, but could be replaced by a single 128-point IDFT upon need. Hence, the LUT number remains constant whereas the binary adder tree grows by one stage. For a large number of subcarriers an optimized, multiplierless IFFT algorithm could well be superior.

An estimation of complexity comparing our optimized IDFT and the IFFT can be done as follows: If the IDFT of Eq. (2.5.14) was performed with the most advanced split-radix fast Fourier transform algorithm [137], the total number of real additions $\alpha(N)$ and multiplications $\mu(N)$ required would be:

$$\alpha(N)=\tfrac{8}{3}N\log_2 N-\tfrac{16}{9}N-\tfrac{2}{9}(-1)^{\log_2 N}+2 \tag{2.5.18}$$

$$\mu(N)=\tfrac{4}{3}N\log_2 N-\tfrac{38}{9}N-\tfrac{2}{9}(-1)^{\log_2 N}+6 \tag{2.5.19}$$

In order to compare our multiplierless, optimized IDFT to the split-radix IFFT we need to relate the number of real multiplications to the number of real additions. Although there are many different implementations of multipliers, generally we can build a multiplier from an adder tree. Typically, to implement a binary multiplier with a resolution of r bit, a number of r adders and r bit shifters are needed. Bit shifters are not counted for this estimate. The required resolution of the adders varies from r to $2r-1$ depending on their position in the tree. In a simple attempt to relate the effort of 1-bit adders representing a multiplier to the number of 1-bit adders representing an r-bit adder therefore we introduce a complexity relation $c(r)$, which for the resolution range $4 \leq r \leq 64$ has been determined for some typical multiplication techniques to be $c(r)\cong 1.5r-1.3$. This way we now can estimate the total of equivalent real adders in an IFFT, namely $\alpha(N)+\mu(N)\,c(r)$, and compare it to the number of real adders $2 \times (2/3)\,(N-5)\,N$ used in our optimized IDFT. (Note: the first factor "2" is due to the fact that complex adders require 2 real add operations. Next, in our technique 4 subcarriers comprise constant pilot tones and therefore require a single add operation only. Last, the DC and Nyquist frequencies do not count either. Therefore we require not N^2 but $(N-5)\,N$ additions only. As a result we find that for $N=32$ ($r \geq 8$), $N=64$ ($r \geq 12$) and for $N=128$ ($r \geq 20$) our IDFT is equivalent or superior to one of the most advanced IFFT algorithms for generating OFDM signals.

The measured overall chip utilization is found to be 84% of slices, 67% of slice registers, and 62% of occupied LUTs based on the FPGA Virtex 5 (XCV5FX200T). Without optimization, we estimate that from the available resources 194% of slices, 187% of slice registers, and 127% of LUTs would be required. Looking only at the part of the design which we optimized in complexity by removing redundancies, we find that in comparison to a non-optimized design we saved 35% of the slices, which fits nicely to the 33% prediction derived from Eq. (2.5.17). In addition, by exploiting redundancies in the storage of modulation coefficients X_k, 85% both of the slice registers and the number of LUT could be saved. This again fits to the prediction for symmetric modulation coefficients X_k, where we calculated savings of 83.5%.

FPGAs make very efficient use of LUTs which allows reducing the computational effort, especially if multiplications can be avoided. In the current design we do not take advantage of the onboard block random access memory (BRAM). However, the LUT contents can very well be stored within the BRAM which provides access cycles as fast as 1.8 ns. Further, a flexible response to changing channel properties can be implemented, if the LUTs are reloaded: LUT contents can be overwritten by loading for example a new set of waveforms via an external interface handled by the on-chip microprocessor. The loading time for these waveforms depends strongly on the implementation. However, once the waveforms are received by the respective LUTs, switching from one LUT to another takes only 5 ns. Updating LUT contents can be achieved during runtime without any loss of data.

4.3.1.4 Hardware Simulations

For designing FPGA-based DSP systems the use of VHDL (very high speed integrated circuit hardware description language) is common. For debugging complex systems, the simulation platform Modelsim verifies the proper functionality of the developed DSP blocks. We feed the debugged Modelsim output to an offline OFDM decoder for analyzing the expected hardware performance. Spectrum and decoded constellation diagram for simulated data are depicted in Fig. 4.28.

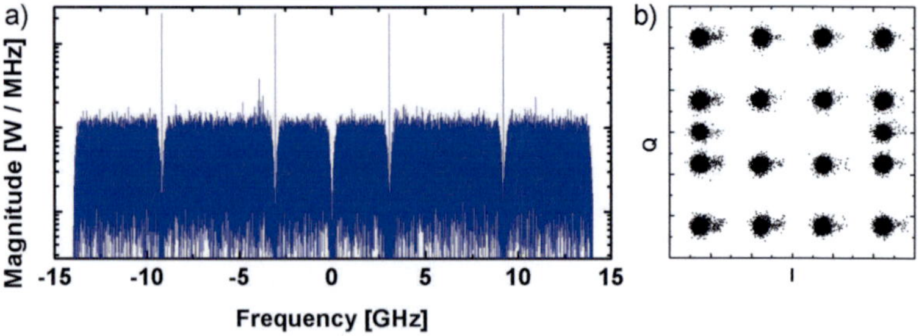

Fig. 4.28 Simulated output spectrum and constellation diagram of all modulated subcarriers and pilot tones showing the proper functionality of the FPGA design. (a) Spectrum with four pilot tones. (b) Constellation diagram. Distortions are due to quantization and clipping noise. A residual EVM of 4.8 % is found for an optimized relation between clipping and quantization noise.

Four spectral lines can be clearly identified as pilot tones since they are not broadened by any modulation. The pilot tones help in carrier phase recovery and frequency offset compensation when decoding the received signal. The overlap of all subcarrier constellation diagrams with the pilot tones is also shown in Fig. 4.28(b). In this simulation distortions are only due to quantization and clipping noise. No analog properties are considered at this point. A residual EVM of 4.8 % is found representing the best signal quality that can be theoretically achieved with the described system [81][82]. This differs from the optimum EVM of 2.7 % specified in the previous section. The discrepancy is due to the differences in arithmetic accuracy: The FPGA uses fixed-point effective 8 bit arithmetic, while EVM = 2.7 % is a limiting value for an "ideal" computation accuracy.

4.3.1.5 Experimental Results

Experimentally, our OFDM Tx works at symbol rates of 437.5 MBd. A number of 58 subcarriers is modulated in the 16QAM format (4 bit / symbol) resulting in an aggregate line-rate of 101.5 Gbit/s. Fig. 4.29(a) displays the electrical spectrum of the measured signal in red and the simulated spectrum in black. The four pilot tones are essential in performing phase recovery and symbol window synchronization. The DC and Nyquist frequency subcarriers remain un-modulated. Together with the 4 pilot tones, we use a total of 58 SC out of 64 for the transport of payload data. Image frequencies arising from digital-to-analog (Tx) and analog-to-digital (Rx) conversions are separated from the signal band by a 437.5 MHz gap. For removing the image frequencies, the slopes of an analog filter must accommodate to this narrow gap — a challenging task, but not out of reach. However, we decided to use a broadband Rx in combination with DSP to suppress the image spectra before further processing. In Fig. 4.29(b) we plot the error vector magnitude (EVM) for each modulated subcarrier.

The measured EVM can be used to reliably estimate bit error ratios of BER $< 10^{-3}$ [115][81][82]. State-of-the-art forward error correction (FEC) algorithms reduce this BER to values below 10^{-9} at the cost of some overhead. Subcarriers far remote from the optical carrier show poorer performance due to a drop in the amplitude transfer function of the transmission system. Fig. 4.29(c) and (d) exemplarily illustrate received constellation diagrams for subcarriers 27 and 51 with an EVM of 7.6 % and 9.7 %, respectively.

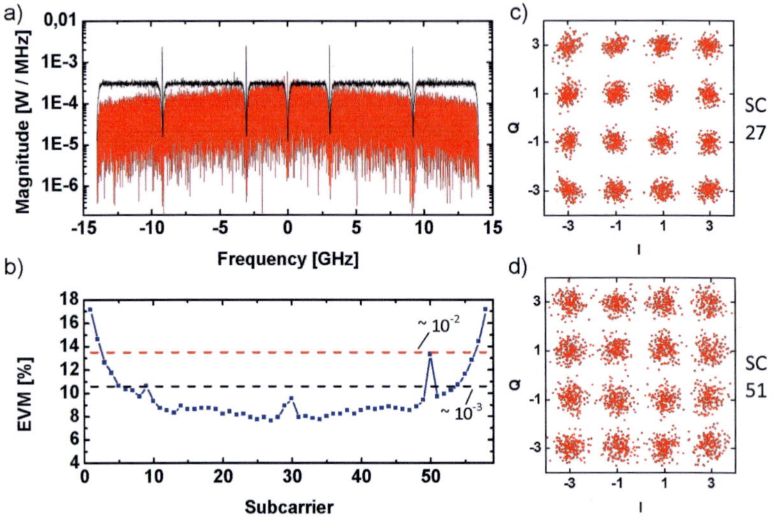

Fig. 4.29 Experimental spectrum, error vector magnitude and constellation diagrams. (a) Received (red) and simulated (black) electrical power spectrum of the transmitted OFDM signal. The simulation is performed for logical signals on the FPGA, and the resulting arbitrary power density is scaled to match the maximum power density of the measured signal. Four pilot tones are used for phase recovery and symbol window synchronization. (b) EVM of modulated subcarriers. Standard limits of the bit error ratio (BER) for forward-error corrected (FEC) "error-free" reception are indicated. (c) Constellation diagram for subcarrier 27, EVM = 7.6 %. (d) Constellation diagram of subcarrier 51 with EVM = 9.7 %.

The DAC's influence on the signal is depicted in Fig. 4.30(a). Each DAC comprises a sample-and-hold stage at the output, which results in a rectangular impulse response $r(t)$ and translates to a sinc-shaped spectrum $R(f)$. The discrete DAC input signal $x_n(t)$ has a spectrum $X_k(f)$ which is periodic. The DAC output spectrum then is $X_k(f)\,R(f)$ (not drawn in Fig. 4.30(b)), so that the spectrum shows a frequency roll-off, red line in Fig. 4.30(b, right). This spectral drop only partially explains the experimental outcome depicted in Fig. 4.29(a), where higher-frequency subcarriers show lesser power. The discrepancy is due to the low-pass character of the system, and the total roll-off equals 7 dB / 13.6 GHz. Still, the average EVM over all subcarriers is 11.1 % which corresponds to a BER of 10^{-3}. Pre-equalization of the system's frequency response can be easily done by changing the pre-computed LUT symbols given that our optimized IDFT algorithm is used for signal generation.

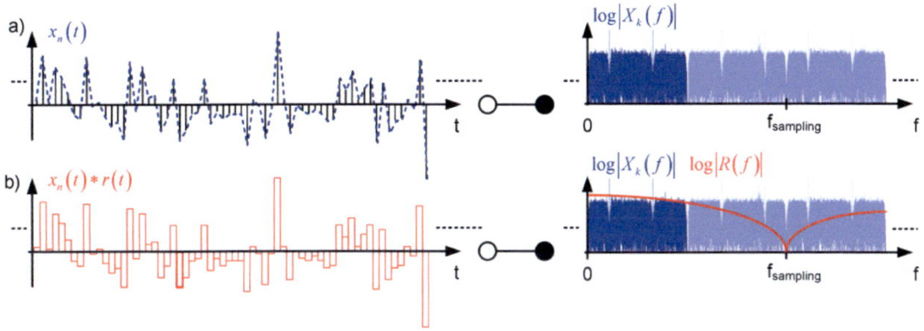

Fig. 4.30 Comparison between an ideal time discrete signal $x_n(t)$ and an ideal DAC output time signal (left) and frequency domain (right). The DAC impulse response is described by a rectangular function $r(t)$ of area 1 and a temporal width equal to the sampling interval. (a) An ideal time discrete signal has a periodic spectrum. Periodic repetitions are in faded blue. A properly chosen anti-aliasing filter blocks these unwanted parts of the spectrum. (b) The ideal DAC output spectrum results from multiplying the (blue) periodic spectrum Fig. 4.30(a) with the (red) sinc-function $R(f)$.

4.3.1.6 Conclusions

We show for the first time a real-time OFDM transmitter achieving line rates of 101.5 Gbit/s. We modulate 58 subcarriers, generated by a 64-point IDFT, with 16QAM. The IDFT uses a reconfigurable and highly optimized algorithm. Four pilot tones help in performing phase recovery and window synchronization at the receiver end. The EVM averaged over all modulated subcarriers amounts to 11.1 %, corresponding to a BER of 10^{-3}. Using state-of-the-art FEC, this error probability level allows error-free reception.

4.3.2 Transmission of 85.4 Gbit/s Real-Time OFDM over 400 km

Generation and transmission of 85.4 Gb/s real-time 16QAM coherent optical OFDM signals over 400 km SSMF with preamble-less reception

R. Bouziane, R. Schmogrow, D. Hillerkuss, P. Milder, C. Koos, W. Freude, J. Leuthold, P. Bayvel, and R. Killey

Optics Express 20(19), pp. 21612–21617 (2012). [J20]

4.3.2.1 Introduction

Orthogonal frequency division multiplexing (OFDM) [23] has attracted much interest within the field of optical communications. Due to its high spectral efficiency and resilience towards fiber dispersion, it has been proposed for a broad range of applications including access and core networks, as well as short links and data centers [23][138][139][140]. While first OFDM experiments were based on offline processing, practical applications call for real-time, highly optimized signal processing. Recently, such OFDM transmitters (Tx) [32][39][38] and receivers (Rx) [35] were demonstrated by employing state-of-the-art field programmable gate arrays (FPGA) along with high speed data converters.

The latest real-time implementations of high speed optical OFDM transmitters include the work published in [39] and [38] where line rates of 101.5 Gb/s and 93.8 Gb/s were reported, respectively. The work presented in [39] used coherent optical OFDM (CO-OFDM) with 16-state quadrature amplitude modulation (16QAM) subcarrier mapping on a single polarization. On the other hand, the results of [38] were achieved with polarization multiplexed CO-OFDM and 4QAM mapping. Both papers considered back-to-back systems without transmission.
This paper describes both the real-time generation and transmission of CO-OFDM signals over 400 km of standard single mode fiber (SSMF) providing more details on a study we previously published in [42]. The real-time transmitter used two FPGAs that encode 85.4 Gb/s of data on 122 subcarriers and a very high speed 128-point inverse fast Fourier transform (IFFT) based on the Spiral Generator [141][142]. A cyclic prefix (CP) with a length of 25 % of the symbol duration is added without the need for additional processing efforts like e.g. multiplexers or first-in first-out buffers (FIFO) enabling dispersion-tolerant transmission over up to 400 km of SSMF. Four pilot tones help to perform frequency offset compensation and carrier phase recovery at the receiver, where data are processed offline. At the receiver, a novel synchronization method is introduced for finding the optimum FFT window position without relying on training symbols. Unlike conventional preamble-based synchronization methods which perform cross-correlations at regular time intervals and let the system run freely in between, the proposed method performs synchronization in a continuous manner ensuring correct symbol alignment at all times.

4.3.2.2 Experimental Setup

The experimental setup comprised a real-time software-defined Tx here operated with OFDM, a recirculating fiber loop, and an Agilent N4391A optical modulation analyzer (OMA) as shown in Fig. 4.31. Two Xilinx Virtex-5 FPGAs (XC5VFX200T) were used to generate the real-time OFDM signal which was then converted into an analog waveform using two 6-bit digital-to-analog converters (DAC) from Micram (operated at 28 GSa/s). The DACs were clocked by an external clock source and used internal programmable clock dividers to provide clocks to the FPGAs. An optical I/Q-modulator encoded the OFDM waveform onto the output of an external cavity laser (ECL) with a center wavelength of 1550 nm and a linewidth of approximately 100 kHz. The resultant signal was amplified using an erbium-doped fiber amplifier (EDFA) and fed to the recirculating loop, which comprised two amplified spans of 50 km SSMF but no dispersion management. The signal was transmitted along with a holding beam provided by a distributed feedback laser (DFB) for maintaining the EDFA at constant operating point. After a pre-determined number of recirculations, the signal was received by the OMA. The OMA is a coherent receiver comprising two 90° optical hybrids (one for each polarization), balanced photodetectors and real-time oscilloscopes (80 GSa/s, 32 GHz electrical bandwidth). A second ECL served as local oscillator for the Rx.

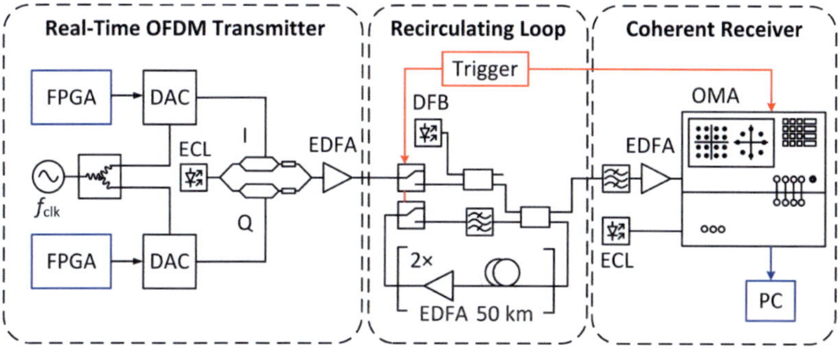

Fig. 4.31 Experimental setup: an FPGA based real-time Tx with optical I/Q-modulator and external cavity laser (ECL). The amplified signal was sent through a recirculating loop comprising two 50 km spans of SSMF with an EDFA. A DFB laser was used as a holding beam in the loop. A trigger source defined the number of roundtrips by controlling optical switches and the optical modulation analyzer (OMA) used to receive the signal. A second ECL served as a local oscillator. Data were processed on an external PC.

The real-time digital signal processing blocks implemented on the FPGA are shown in Fig. 4.32. A $2^{15} - 1$ pseudo-random bit sequence generation block (PRBS) generated the data and fed them to a QAM mapper, which mapped each 4 bits into a 16QAM symbol. The PRBS generators on the two FPGAs were synchronized. The IFFT block, which was created using the Spiral Generator framework [142], processed 128 complex inputs to produce the time domain OFDM signal with 10 bits of arithmetic precision. The input to the IFFT included 122 data subcarriers and 4 evenly-spaced pilot tones. The pilot tones were used to perform frequency offset compensation and carrier phase recovery at the receiver. The DC and Nyquist subcarriers (with indices 0 and 64, respectively) were left blank and thus carried no data. The output of the IFFT was clipped to reduce the peak-to-average power ratio (PAPR) of the signal and, hence, reduce the quantization noise in the DAC. This was then followed by a CP insertion stage. The data are passed from the FPGA to the DAC through 24 multi-gigabit transceivers (GTX) operating in parallel which corresponds to four 6-bit samples. The GTXs act as on-chip multiplexers where parallel registers are serialized [143]. The width of these registers was set to 32 bits in previous works, e.g. [39]. In this work, the registers were set to be 40 bits wide. In this way, $4 \times 40 = 160$ samples (rather than $4 \times 32 = 128$ samples) were processed within each clock cycle. This corresponds to the 128 point IFFT output plus an additional 32 samples (25 %) of cyclic prefix. To enable this operation, the clock divider on the DAC was reconfigured to match the new register widths. With this technique, no additional resources on the FPGA needed to be utilized for CP insertion.

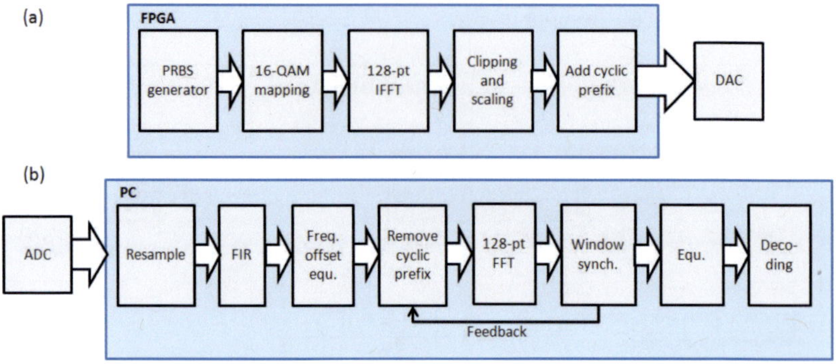

Fig. 4.32 (a) Real-time FPGA processing blocks of transmitter comprising PRBS generation, 16QAM mapping, an IFFT, clipping and scaling, and cyclic prefix insertion. (b) The receiver offline processing blocks comprise resampling, FIR filtering, frequency offset compensation, cyclic prefix removal, an FFT, window synchronization, equalization (Equ.), and decoding of the data. The signal quality is evaluated through EVM measurements.

The receiver signal processing was carried out offline using Matlab and a block diagram of the main components is shown in Fig. 4.32. Standard OFDM processing blocks including finite impulse response (FIR) filtering to compensate for the roll-off in the system frequency response, frequency offset compensation, CP removal, FFT, and equalization for channel estimation and dispersion compensation were used. The symbol synchronization however, was achieved using a novel method without relying on a preamble (training symbols). The next section will explain this in more detail.

4.3.2.3 Transmission experiment and results

The system was tested in the optical back-to-back configuration whereby the recirculating loop (shown in Fig. 4.31) was bypassed and the output of the transmitter was fed to the receiver directly. The ensemble-averaged 28 GHz spectrum of the OFDM signal is shown in Fig. 4.33(a). Four pilot tones broadened by the CP can be seen. The roll-off stems from the frequency response of the system and is equalized in the Rx. The signal quality was assessed using error vector magnitude (EVM) [81][82] and the EVM performance for each subcarrier k is depicted in Fig. 4.33(b). Most subcarriers had an EVM equal to or better than 10%. Subcarriers close to $k = 61$ suffered higher EVM than the rest because they are significantly affected by the optical carrier noise which is located close to these subcarriers. Subcarriers close to $k = 0$ and $k = 121$ suffer from the system roll-off, as they are located at high frequencies. A clear back-to-back received constellation diagram for subcarrier $k = 5$ is shown in Fig. 4.33(c).

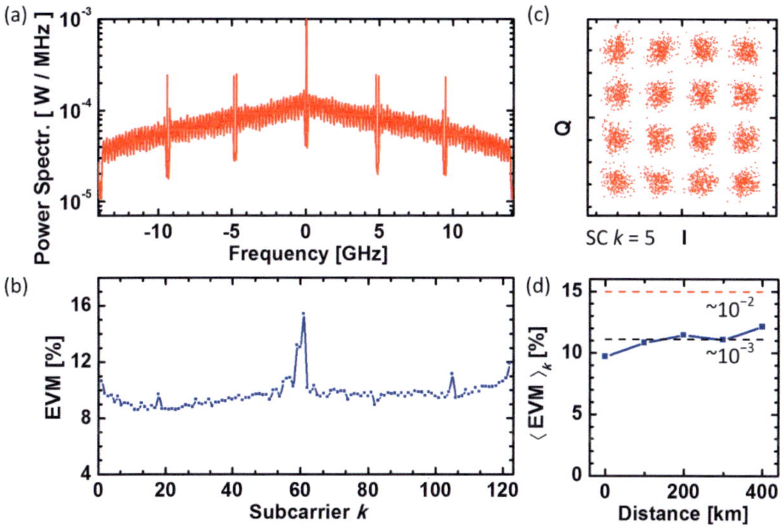

Fig. 4.33 Experimental results. (a) Measured and ensemble-averaged spectrum. Four pilot tones broadened by the CP are seen. (b) Signal quality, expressed by EVM for different subcarrier positions k in a back-to-back measurement. (c) Constellation diagram of subcarrier $k = 5$ and back-to-back configuration. (d) Average EVM for all subcarriers after the transmission distance. Values of EVM corresponding to BER of 10^{-3} and 10^{-2} are indicated.

Next, the transmission experiment was carried out over several transmission distances and the average EVM over all 122 data subcarriers was measured in each case. Fig. 4.33(d) presents the average EVM as a function of the transmission distance up to 400 km. The figure also indicates the values of EVM corresponding to bit error ratios (BER) of 10^{-3} and 10^{-2}. As can be seen from the figure, the system achieved acceptable EVM for all distances (up to 400 km).

The output of the FFT window synchronization module can be seen in Fig. 4.34 for both the back-to-back configuration and after 400 km transmission distance. In the back-to-back measurements, suitable windows can be found within the range of the CP of 25 % of the symbol length, i.e., within 1 – (1 / 1.25) = 20 % (1.26 rad) of the maximum phase difference 2π. This is the trough width in Fig. 4.34(a). After transmission over 400 km of fiber without dispersion management, the full length of the CP is needed for dispersion compensation. Therefore, only a fixed FFT window position at the minimum $\langle\sigma(|C_k|)\rangle_k$ offers best performance (cf. Fig. 4.34(b)).

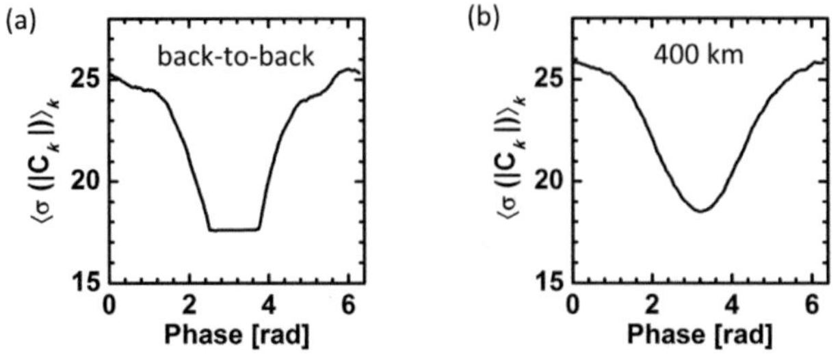

Fig. 4.34 Output of the synchronization module (a) in the back-to-back configuration and (b) after 400 km transmission.

4.3.2.4 Conclusions

This paper presented the generation and transmission of real-time 16QAM coherent optical OFDM signals. The FPGA-based transmitter ran at 28 GSa/s and achieved a data rate of 85.4 Gb/s. A 25 % cyclic prefix was added without the need for additional FPGA processing resources and enabled dispersion-tolerant transmission over up to 400 km of SSMF. A key aspect of the work was the development and demonstration of a novel method for OFDM symbol synchronization without relying on training symbols. Unlike conventional preamble-based synchronization methods which perform cross-correlation at regular time intervals and let the system run freely in between, the proposed method performs synchronization in a continuous manner ensuring correct symbols are used all the time.

4.4 Real-Time OFDM vs. Nyquist Pulse Shaping

After the introduction of real-time Nyquist pulse-shaping and OFDM signal generation in the previous sections, the two techniques are compared with respect to resource requirements and signal performance.

Real-time OFDM or Nyquist pulse generation – which performs better with limited resources?

R. Schmogrow, R. Bouziane, M. Meyer, P. Milder, P. Schindler, R. Killey, P. Bayvel, C. Koos, W. Freude, and J. Leuthold

Opt. Express 20(26), pp. B543–B551 (2012). [J16]

Digital signal processing (DSP) resources ultimately limit the maximum speed of real-time multiplexing techniques such as orthogonal frequency division multiplexing (OFDM) [23] or Nyquist wavelength division multiplexing (N-WDM) [24]. These two multiplexing methods are attractive, because in theory they both offer a spectral efficiency (SE) close to or at the Nyquist limit [79]. Yet, despite their similarities [79], where OFDM is mathematically described by a superposition of sinc-shaped subcarriers, while N-WDM is made up of temporal sinc-pulses, the actual implementation and thus the DSP resource requirements are quite different depending on the selected processing technique. The question remains: Which of the two techniques can be implemented more efficiently with respect to DSP resources?

Spectrally efficient ultra-dense WDM networks call for optimum channel utilization without inter-channel interference (ICI). At best, the channel spectra should be of rectangular shape. This way, independent channels can be added without overlap. This enables independent channels which need not be synchronized and use free-running independent laser sources as optical carriers. Due to their potentially high SE, both Nyquist pulse shaping and multiband OFDM [144] are promising options. Spectra for these two multiplexing techniques are schematically depicted in Fig. 4.35. We briefly discuss the concepts and focus on the relevant parameters for generating nearly-rectangular spectra with either of the schemes.

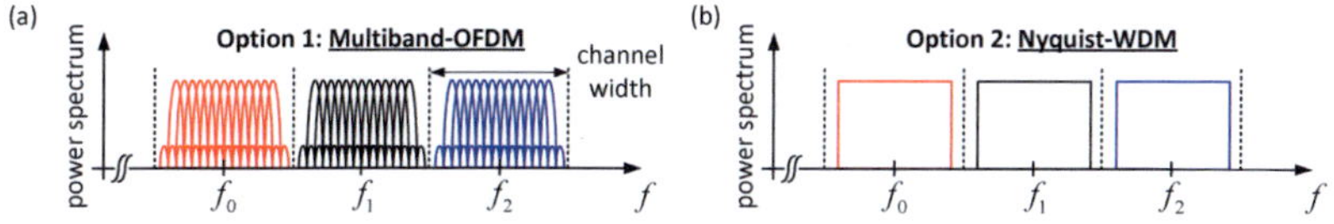

Fig. 4.35 Pulse shaping for ultra-dense WDM with enhanced SE. Independent and ICI-free channels are desired. (**a**) **Multiband OFDM:** Each channel comprises several electrically generated subcarriers. (**b**) **Nyquist-WDM:** Sinc-shaped pulses with rectangular spectrum. Both techniques efficiently use the available spectrum.

Option 1 – OFDM: Electrically generated OFDM signals comprising multiple subcarriers (SC) can be transmitted in independent WDM channels, Fig. 4.35(a). With an increasing number of SCs per channel within a fixed bandwidth, the OFDM spectrum approaches a rectangle. Therefore, by a proper choice of SC number, inter-channel guard-bands can be reduced without introducing ICI.

Option 2 – N-WDM: Electrically generated Nyquist pulses are shaped with so-called finite duration impulse response (FIR) filters, have nearly rectangular spectra, and are transmitted in independent WDM channels (N-WDM), Fig. 4.35(b). The higher the filter order R is, the closer the spectrum approaches a rectangle. If R is chosen appropriately, then inter-channel guard-bands can be kept small without introducing ICI.

Several real-time transmitter (Tx) concepts producing digitally generated OFDM and Nyquist pulses have been demonstrated. Here we confine the discussion to implementations with the potential of generating 100 Gbit/s using a dual-polarization scheme. For real-time OFDM, recent publications demonstrated data rates of 101.5 Gbit/s [39] and 85.4 Gbit/s [43] in a single polarization and 93.8 Gbit/s [38] in a polarization divison multiplexed (PDM) [78] configuration. So far, real-time Nyquist sinc-pulse Tx were demonstrated at speeds of 112 Gbit/s [79] and 150 Gbit/s [41], both using PDM.

In this paper we compare the system performance and DSP resource requirements of OFDM and Nyquist sinc-pulse generation for two selected DSP approaches, exemplarily. This work is an in-depth discussion of our work previously published in [145] extended by simulations of the implemented DSP and receiver (Rx) discussions. First, the implemented code to be run on field programmable gate arrays (FPGA) is evaluated in terms of achievable performance and resource utilization. Second, both a single-channel and an ultra-dense WDM scenario with three carriers are experimentally assessed, using two software-defined Tx to generate either OFDM with 80 SCs or Nyquist sinc-pulses formed with FIR filters of order $R = 32$, 64, and 128. With both multiplexing techniques we transmitted either the format quadrature phase shift keying (QPSK) or we used quadrature amplitude modulation with 16 states (16QAM). Finally, we briefly discuss potential OFDM and Nyquist pulse Rx.

4.4.1 VHDL/Verilog Simulations and Resource Requirements

First we investigated the potential performance of the two pulse shaping techniques in terms of signal quality and spectral shape using the target field programmable gate arrays (FPGA) that were to be used in the experiments. To this end, the designs developed with the hardware description languages Verilog (for the inverse fast Fourier transform only, IFFT) and VHDL (very high speed integrated circuit hardware description language) were investigated on the register transfer level (RTL), a design abstraction which models the signal flow of synchronous digital circuits. The simulations (performed with the Modelsim software) used a test bench (Xilinx ISE software) for obtaining the bit-level-accurate outputs to be expected from the FPGAs. Additionally, we determined the resource requirements of each DSP design. The DSP blocks are pictured in Fig. 4.36(a) (OFDM) and in Fig. 4.36(b) (N-WDM), respectively.

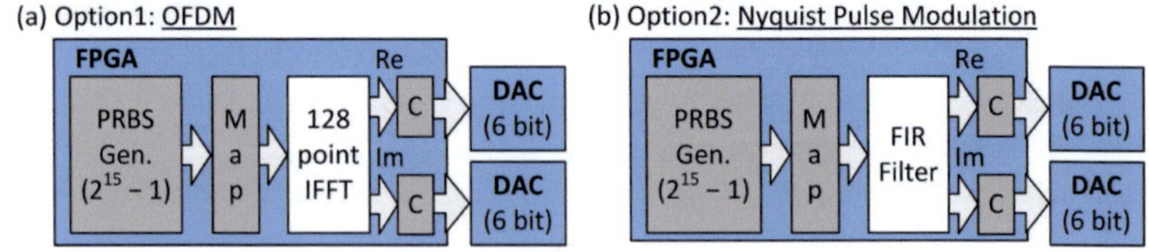

Fig. 4.36 Digital signal processing blocks (DSP) implemented on field programmable gate arrays (FPGA) for (**a**) real-time OFDM and for (**b**) Nyquist sinc-pulse shaping. A pseudo random binary sequence (PRBS) generator emulates payload data. A mapper (Map) translates binary data to QPSK/QAM symbols. The digital pulse shaping is either performed by an IFFT (OFDM) or by an FIR filter (Nyquist). Clipping blocks (C) limit the signal amplitudes to reduce the peak-to-average power ratio. Finally, 6-bit digital-to-analog converters (DAC) generate the output waveforms for subsequent IQ-modulation of the optical carrier.

Option 1 – OFDM: The DSP blocks used for real-time OFDM generation are depicted in Fig. 4.36(a). A pseudo random binary sequence (PRBS) generator produced binary data (periodicity: $2^{15} - 1$) that served as input. Following this, bit mapping to the QPSK/QAM states was performed. In the next processing block, a 128 point IFFT based on a mixed-radix Cooley-Tukey fast Fourier transform (FFT, radices 16 and 8, generated with the Spiral tool) [142] produced a complex-valued OFDM waveform with 80 SCs carrying either QPSK or 16QAM data. Four additional pilot tones were used for equalization. The remaining SCs were set to zero leading to an oversampling factor of $128 / 85 \approx 1.5$. Oversampling is needed to remove the image spectra produced by digital-to-analog converters (DAC) with realizable low-pass filters. The generated signals had an optical bandwidth close to 13 GHz carrying either 25 Gbit/s (QPSK) or 50 Gbit/s (16QAM) of data. The computational precision of the IFFT was 10 bit and the output waveform was quantized and clipped to 6 bit according to the DAC resolution of the target system. Clipping reduces the peak-to-average power ratio (PAPR) of the signals [23]. The simulated OFDM spectrum (assuming an ideal 6 bit DAC) is depicted in Fig. 4.37(a). The frequency is normalized to the sampling rate F_s (in the experiments $F_s = 20$ GSa/s) of the used DACs. Four pilot tones can be identified as spectral lines in Fig. 4.37(a), since their amplitudes were kept constant throughout all OFDM symbols. The simulated constellation diagram for 16QAM modulated SCs is shown in Fig. 4.37(b). In this diagram all 80 SCs contribute. It can be seen, that some of the SCs produce constellation points with an offset to the lower left. This has been compensated for at the receiver (Rx). The simulations predicted an achievable error vector magnitude (EVM) [81][82] of 4.2 %. The residual error is due to limited computational precision, clipping and quantization noise.

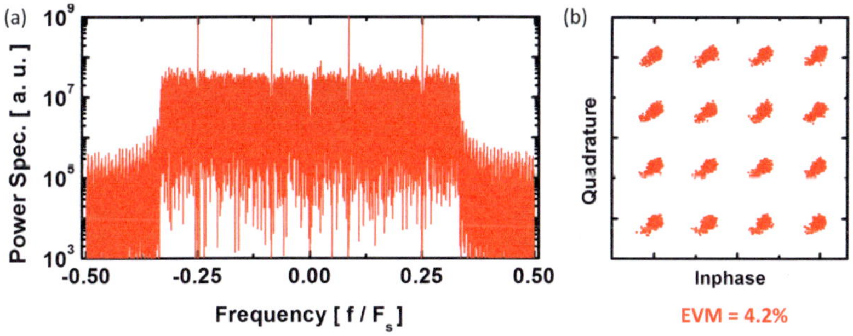

Fig. 4.37 Spectrum and best achievable constellation diagram for simulated 16QAM OFDM. (**a**) Spectrum comprising 80 SCs used for data transmission and 4 pilot tones. (**b**) Constellation diagram comprising data of all 80 SCs. The minimum achievable EVM is 4.2%.

Option 2 – Nyquist sinc-pulse generation: The DSP blocks used for real-time Nyquist sinc-pulse shaping are depicted in Fig. 4.36(b), where the same PRBS generator was used and the binary data was mapped to QPSK/16QAM symbols. Instead of using an

IFFT, this time QPSK/16QAM data fed an FIR filter which processed 128 samples in parallel within each clock cycle. The impulse response of this filter was sinc-shaped, resulting in a near-rectangular signal spectrum. For Nyquist pulse shaping we used two-fold oversampling, i.e., two samples were generated for each transmitted symbol. Due to the use of look-up tables (LUT) [79], no multiplications during runtime were needed, a significant advantage when it comes to processing complexity and resource requirements. FIR filters of orders R = 32, 64, and 128 were implemented. As expected, the spectrum evolved towards an ideal rectangle with increasing filter order. The computational precision was dynamically adjusted according to [41], leading to an effective precision of 9.9 bit. Again a clipping module lowered the PAPR by removing amplitude peaks for quantizing the output waveform with 6 bit. Spectra obtained from the simulations of the VHDL designs using random input data are depicted for a filter order R = 32 (Fig. 4.38(a)) and for a filter order R = 128 (Fig. 4.38(b)), respectively. Due to two-fold oversampling, the spectra are confined to the region $|f / F_s| \leq 1 / 4$. Even the lower-order R = 32 filter shows a good out-of-band suppression and would be expected to perform well in ultra-dense WDM experiments. The generated Nyquist signals had a bandwidth close to 10 GHz for a symbol rate of 10 GBd. The data rates were either 20 Gbit/s (QPSK) or 40 Gbit/s (16QAM). Unlike OFDM DSP, Nyquist pulse shaping with 2^s-fold oversampling (integer s) shows no residual EVM. This is due to the fact that each pulse maximum coincides with zeros of all other pulses, independent of the filter order that has been implemented.

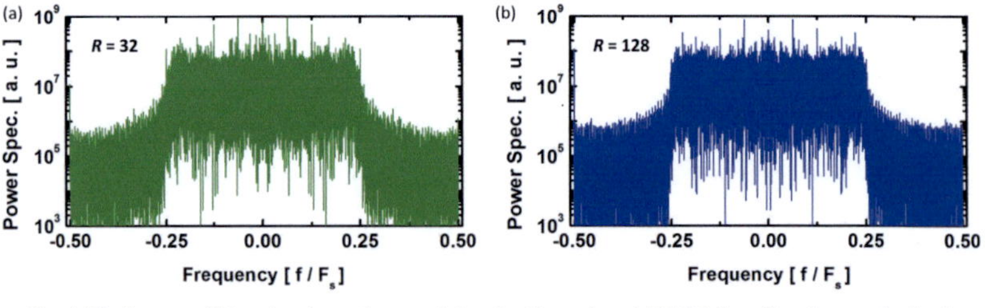

Fig. 4.38 Spectra of Nyquist sinc-pulses modulated with random 16QAM data. Results are obtained by VHDL simulations. Spectra shaped with an FIR filter (**a**) of order R = 32 and (**b**) of order R = 128.

Resource requirements: Table 4.3 summarizes the necessary FPGA resources for implementing the OFDM and N-WDM Tx. All numbers are based on synthesis results using the Xilinx XC5VFX200T FPGA. We compare the required resources for generating a complex-amplitude OFDM or Nyquist sinc-pulse waveform and conclude: While the number of slice registers and slice LUTs break even for the OFDM and Nyquist signals with filter orders around 64, our LUT-based Nyquist pulse shaping technique removes the need for any DSP48E slices as they support multiple functions (e.g. resource-hungry multiplications).

Table 4.3 FPGA resource utilization for complex signal generation

Resource	**OFDM**		**Nyquist pulse shaping**		
	QPSK	16QAM	QPSK/16QAM		
	128-point IFFT		**32nd order**	**64th order**	**128th order**
Slice registers	39850	41790	28630	44840	74230
Slice LUT	46720	50508	19910	36790	64410
DSP48E (e.g. multipliers)	380	380	0	0	0

4.4.2 Experimental Setup

The experimental setup is depicted in Fig. 4.39. Three single-polarization WDM channels were generated using tunable external cavity lasers (ECLs) and a pair of nested $LiNbO_3$ Mach-Zehnder modulators (MZM), which were connected in a configuration such that the two outer channels were modulated independently from the center channel. We employed two software-defined transmitters [73] each based on pairs of FPGAs (Xilinx Virtex 5 XC5VFX200T) and high-speed Micram DACs with 6 bit resolution. The DACs operated at 20 GSa/s and were followed by anti-aliasing filters with 12.5 GHz bandwidth. These transmitters generated the modulator drive waveforms (Tx I for the center channel, Tx II for the two outer channels in Fig. 4.39). A $2^{15} - 1$ PRBS was generated on the FPGAs and transmitted. The polarization states of the WDM channels were adjusted to obtain the worst-case ICI penalty. An optical spectrum analyzer (OSA) measured the WDM signal spectra and the optical signal-to-noise ratio (OSNR) for single channel configuration. The OSNR and therefore the relative amplified spontaneous emission (ASE) noise loading were adjusted with a variable optical attenuator (VOA). The center channel was selected with an optical filter, coherently detected using an Agilent optical modulation analyzer (OSA), and the data recovered using off-line DSP.

In order to obtain an equal data rate of 20 Gbit/s (QPSK) and 40 Gbit/s (16QAM) for the OFDM and Nyquist-WDM Tx signals, the OFDM signal should have been sampled with 16 GSa/s whereas for Nyquist-WDM a sampling rate of 20 GSa/s would have been appropriate. This stems from the fact that the Nyquist-WDM DSP uses two-fold oversampling and the OFDM DSP oversamples by a factor of $128 / 85 \approx 1.5$. However, the available anti-alias low-pass filters with a cutoff-frequency of 12.5 GHz would not have removed the whole of the image spectra at 16 GSa/s DAC operation. For this reason we have chosen 20 GSa/s for OFDM. In order to guarantee a fair comparison (independent of signal bandwidth) between OFDM and Nyquist signaling, we derived the actual SNR from the measured $OSNR_{0.1nm}$ [120].

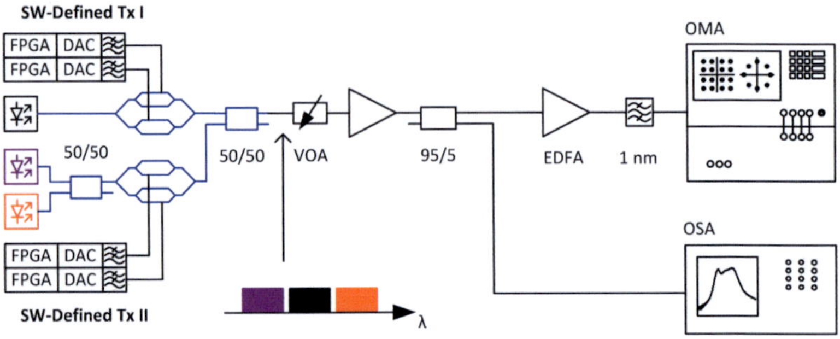

Fig. 4.39 Experimental setup for real-time OFDM and Nyquist pulse shaping measurements. Two software-defined transmitters generate either complex Nyquist sinc-pulses or OFDM data to be modulated onto three free-running ECL. The blue signal path marks polarization maintaining optics. A variable optical attenuator (VOA) varies the signal power before the first EDFA hence varying the OSNR. A standard 1 nm optical filter removes ASE away from the three channels. A second EDFA guarantees optimum input power to the Agilent optical modulation analyzer (OMA) where the signals are coherently received and processed offline.

4.4.3 Experimental Results

4.4.3.1 Single Channel Characterization

In a first experiment, we tested the signal-to-noise ratio (SNR) sensitivity of single channel OFDM signals generated by the 128 point IFFT and compared the results with those obtained with Nyquist pulse-shaped signals generated with FIR filters of orders R = 32, 64 and 128.

The results of the EVM [81][82] measurements as a function of SNR are plotted in Fig. 4.40. For OFDM we take the average EVM of all 80 SCs. It can be seen that the quality of the Nyquist signal is independent of the filter order as was expected from the simulations. Furthermore, it can be concluded that in the case of QPSK modulation, both OFDM and Nyquist pulse shaping show very similar SNR performance, see Fig. 4.40(a).

For 16QAM OFDM, however, the EVM was found to be slightly worse than for Nyquist-WDM, see Fig. 4.40(b). We attribute this increased implementation penalty in part to the 10 bit precision of the IFFT core, which was limited by the available resources on the FPGAs. Another limitation was imposed by the finite DAC resolution of 6 bits. Both limitations more seriously affect high order modulation formats.

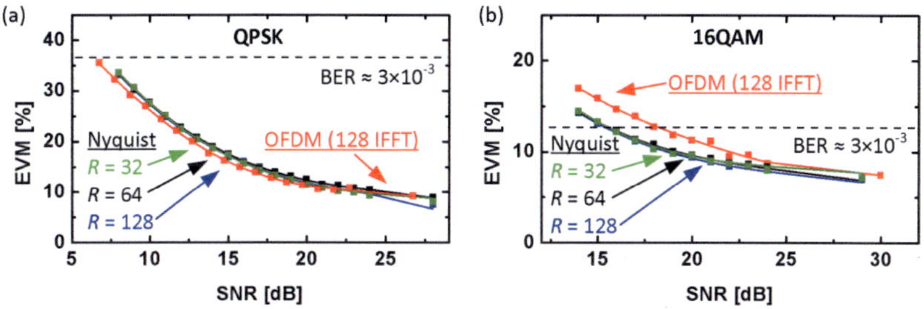

Fig. 4.40 EVM as a function of SNR for OFDM and Nyquist pulse shaping. Dashed lines mark a bit error ratio (BER) limit of 3×10^{-3}. (**a**) QPSK encoded Nyquist sinc-pulses generated with different filter orders R (and different roll-off) perform equally well. OFDM transmission with QPSK modulation shows comparable quality as Nyquist pulse shaping. (**b**) For 16QAM, OFDM suffers from slightly higher implementation penalties than Nyquist pulse transmission.

Constellation diagrams are plotted for 16QAM Nyquist pulse shaping (Fig. 4.41(a), blue) and SC 26 of the OFDM signal (Fig. 4.41(b), red) in back-to-back configuration. In Fig. 4.41(c) the EVM as a function of SC index is plotted. Subcarriers far away from the optical carrier are degraded due to a decreased SNR caused by the roll-off of the analog electronics. This could have been compensated for by pre-emphasizing SCs with lower SNR. However, in doing so, the power of other SCs reduces, so that the average SNR throughout all SCs remains constant.

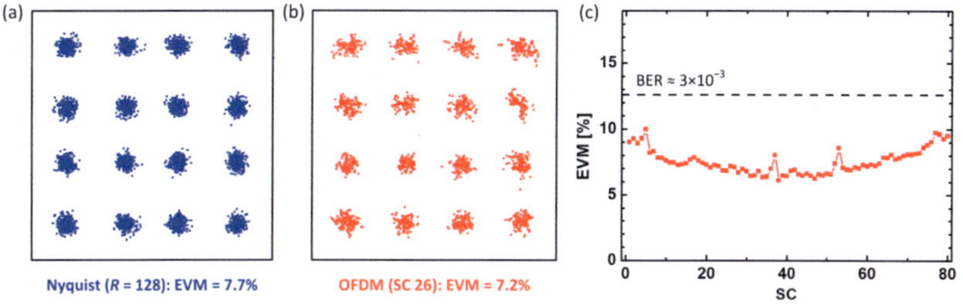

Fig. 4.41 Back-to-back received constellation diagrams for 16QAM Nyquist pulse shaping and OFDM along with the EVM as a function of OFDM SC. (**a**) 16QAM Nyquist constellation diagram shows an EVM of 7.7 %. (**b**) 16QAM OFDM constellation diagram of the 26th SC with an EVM of 7.2%. (**c**) EVM as a function of subcarrier number. SCs far away from the center show an increased EVM. The horizontal dashed line specifies a BER limit of 3×10^{-3} for hard decision forward error correction.

4.4.3.2 Multi Channel Characterization

In order to assess the achievable SE we tested all designs in an ultra-dense WDM experiment comprising three free-running carriers. Since only the linear inter-channel crosstalk affecting the middle channel was investigated, three channels were sufficient as little additional information (for linear operation) can be gained from increasing the number of channels beyond this. The OSNR was set to its maximum achievable value. With the wavelength of the center channel kept constant, the outer channel wavelengths were varied to adjust the SE of the WDM signal. The guard-band between the edges of the signal bands was varied from 10 GHz down to zero for Nyquist-spacing. Fig. 4.42 shows the WDM spectra with 5 GHz and 500 MHz guard bands.

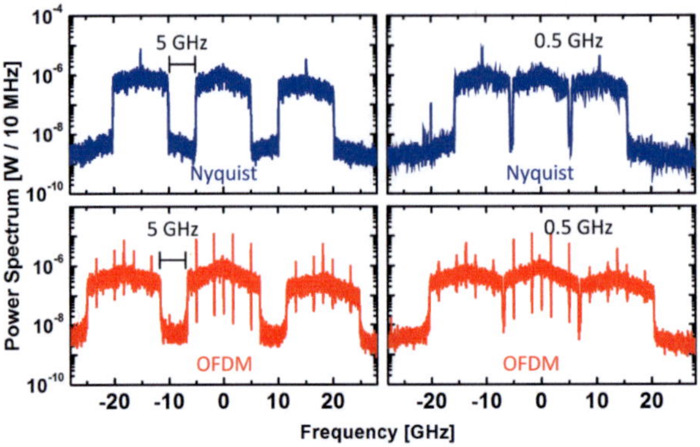

Fig. 4.42 Three-channel WDM spectra with guard-bands of 5 GHz (left) and 500 MHz (right) for Nyquist pulse shaping with filter order $R = 128$ (top, blue) and OFDM (bottom, red).

To investigate the potential SE of both pulse shaping techniques in ultra-dense WDM networks, we measured the EVM penalty as a function of inter-channel guard-band width. The results for QPSK modulation are depicted in Fig. 4.43(a). For guard-bands > 1 GHz there was virtually no influence of the Nyquist pulse-shapers' filter order R on the signal quality. The ICI was negligible. For very small guard-bands (< 1 GHz), high filter orders perform slightly better. At some point, however, severe crosstalk is observed. This stems mostly from the fact that we use independent, free-running lasers whose center frequencies are not perfectly stable. Despite this, we can conclude that, even for small guard-bands, low Nyquist filter orders R suffice.

Similarly to Nyquist-WDM, good performance is also observed with the OFDM transmitters, even with small guard-bands between channels. The OFDM signal quality degrades gradually for guard-bands < 3 GHz which can be explained by the lower out-of-band suppression of the OFDM spectrum as compared to that of the Nyquist filtered signals. To reduce this effect, a larger number of SCs would be needed to make the signal spectrum more rectangular. The lesser increase in EVM penalty with reducing guard-band width for OFDM, compared to that observed with Nyquist pulse shaping can be explained by the outermost SCs of the OFDM signal being affected by ICI. At some point they can no longer be used for data transmission, while the inner-SC EVM remains low. The results for 16QAM modulation are shown in Fig. 4.43(b). These results follow the observations made in the QPSK measurements.

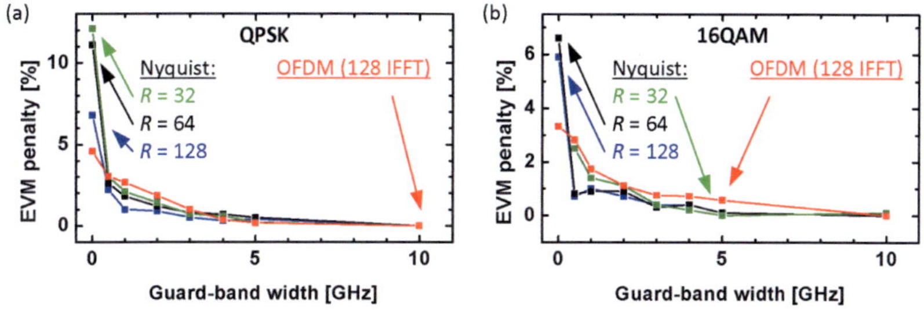

Fig. 4.43 Measured EVM penalty as a function of guard-band width to assess the potential SE of Nyquist pulse shaping and multiband OFDM. (**a**) Results for QPSK modulation show that even Nyquist pulse-shapers with low order *R* perform well. For OFDM, ICI occurs for guard-band widths < 3 GHz due to the smaller out-of-band suppression compared to Nyquist filtered signals. (**b**) Results for 16QAM confirm observations made for QPSK.

4.4.4 Discussion on Rx complexity for OFDM and Nyquist pulse shaping

So far, only a few real-time optical receivers (Rx) for OFDM [144][36][35] and single carrier *M*-ary QAM [128] have been demonstrated. Due to the increased complexity of the Rx as compared to the Tx, achieved data rates are significantly lower than for real-time Tx [79][39][43][38].

The main difference between a multicarrier OFDM Rx and a single carrier Nyquist pulse Rx is the process of data extraction and equalization. An OFDM Rx comprises an FFT which demultiplexes the orthogonal SCs [23]. While the OFDM signal can be made tolerant to chromatic dispersion (CD) and polarization mode dispersion (PMD) through the addition of a cyclic extension [23], which can be achieved with low complexity DSP, this comes at the cost of increased overhead and thus reduced SE. It is preferable in high SE systems to employ FIR filters to equalize CD and PMD, avoiding the need for a large cyclic extension. Likewise, for Nyquist pulse Rx, FIR filters are needed to compensate for CD and PMD, and also to effectively remove spectral components from neighboring channels. An extensive analysis of receiver complexity for single carrier (e.g. Nyquist pulse modulation) and multi carrier (OFDM) systems is given in [146]. In general, high order FIR filters are more effectively realized in the frequency domain using FFTs by employing frequency domain equalization (FDE) [146].

Analog electrical low-pass filters at the Rx can also contribute to the removal of unwanted components of the neighboring WDM channels for both multiband OFDM and Nyquist pulse transmission. Such analog filters relax the requirement on the Nyquist pulse Rx digital filter significantly. The non-flat frequency response of the transceiver electronics along with accumulated CD can be pre-compensated at the Tx [65], hence further reducing processing efforts for single carrier Rx.

4.4.5 Conclusions

We directly compared, for the first time, performance and DSP resource requirements of FPGA based OFDM and Nyquist pulse shaping. Therefore two real-time implementations were compared, namely Nyquist pulse shaping with LUTs [39] and OFDM generation using the Spiral IFFT core [142]. There are several studies on DSP complexity from a theoretical point of view [146] concluding that OFDM and Nyquist pulse shaping are similarly complex. An actual implementation using state-of-the-art FPGAs and DACs, however, reveals the strengths and weaknesses of each technique from a practical point of view. For instance, the LUT based Nyquist pulse shaping approach offers the possibility to compensate for the frequency response of the hardware by simply changing the pre-computed filter responses stored in the LUTs whereas for OFDM, the implemented IFFT core was optimized with respect to the specific 16QAM SC modulation format. Hence it did not allow for the OFDM typical one-tap equalizer [23] to be made use of in this configuration. As future work, real-time OFDM and Nyquist pulse Rx ought to be compared.

Experiments were performed for a single wavelength and in an ultra-dense WDM scenario. Both, OFDM and Nyquist-WDM channels can be placed next to each other close to the Nyquist limit without experiencing significant inter-channel crosstalk, even though free-running lasers provided the optical carriers. However, the Nyquist sinc-pulses offer similar or slightly better performance than OFDM for very small guard-band widths (over the range 0.5–1 GHz). This can be explained by better out-of-band suppression of the N-WDM signals which calls for an increase in the number of SCs for OFDM. The overall OFDM signal quality in our experiments was essentially determined by the limited precision of the implemented IFFT and the limited effective resolution of the DACs. For Nyquist pulse shaping, an increase in filter order beyond R = 128 is not expected to further significantly minimize the crosstalk. At small guard-bands (i.e. lower than 10% of the channel spacing), even relatively low order (e.g. 32nd order) low-resource Nyquist pulse shaping suffices.

5 Systems

In this chapter several system experiments are presented employing Nyquist and OFDM signaling in several variants. This includes filtering techniques for Nyquist signaling, setting new spectral efficiency (SE) marks, and employing Nyquist and OFDM in an access network.

5.1 Nyquist Pulse-Shaping

Nyquist signals can be generated by different means. First generation of Nyqust pulses with digital, optical, and analog electrical filters is discussed and the performance is compared. Then it is shown how digitally shaped Nyquist signals can be used to show SE way beyond 10 bit/s/Hz.

5.1.1 Pulse-Shaping with Digital, Optical and Electrical Filters

Pulse-Shaping with Digital, Electrical, and Optical Filters – A Comparison

R. Schmogrow, S. Ben-Ezra, P.C. Schindler, B. Nebendahl, C. Koos, W. Freude, and J. Leuthold

J. Lightw. Technol. 31(15), pp. 2570–2577 (2013).

Shaping the pulse envelope of M-ary quadrature amplitude modulated (QAM) signals has attracted quite some attention recently. Pulse-shaping techniques allow for instance a decrease of the channel spacing, and with this an improvement of the spectral efficiency (SE) of wavelength division multiplexed (WDM) transmission, while still avoiding linear crosstalk among neighboring channels. Pulse-shaping may also improve the nonlinear transmission performance.

Among the many possible pulse-shapes, sinc-shaped pulses with a corresponding rectangular spectrum are of particular interest as they allow transmission in the Nyquist WDM regime. For Nyquist WDM, channel spacing and symbol rate are identical. Sinc-shaped Nyquist pulses are also special in a way that their pulse form meets the Nyquist criterion, according to which impulse maxima coincide with the zeros of neighboring pulses, so that inter-symbol interference (ISI) is avoided. In practice, shaping of a pulse can be achieved by performing pulse-shaping in the digital, in the electrical, or optical domain using appropriate filters. While at lower symbol rates (< 35 GBd) all shaping techniques are available, at higher symbol rates only analog electrical or optical techniques are at hand. Each of the methods has advantages and disadvantages. Unfortunately, the various methods have never been directly compared.

In this paper we compare digital, electrical and optical pulse-shaping techniques. The comparison is performed for lower symbol rates where all techniques can be implemented. More precisely, we form a sinc-shaped pulse from a 20 GBd quadrature phase shift keyed (QPSK) signal and investigate the influence of the shaping technology on the signal quality for a single optical carrier. Sinc-shaping in the digital domain is performed by our software-defined transmitter (Tx). This allows us to create an almost perfect sinc-shaped pulse form with virtually zero roll-off. Electrical and optical filters are alternatively used to approximate sinc-shaped pulses by analog means. Last, we assess the performance of the signals in a setup with three carriers where the channel spacing of 20 GBd QPSK signals is varied from 17 GHz to 50 GHz. This way, we explore pulse-shaping for the sub-Nyquist WDM, the Nyquist WDM, and the ultra-dense WDM regimes. To make the comparison as fair as possible, we have optimized not only the Tx but also the receiver (Rx) with a sophisticated equalization technique based on Nyquist brick-wall filtering to minimize the ISI for each of the transmitters. It is found that digitally formed sinc-shaped pulses provide superior performance at low symbol rates. However, sinc-shaped pulses generated by electrical filters are not so far off, and pulses shaped by optical filters still outperform plain unshaped signals.

5.1.1.1 Digital, Electrical or Optical Pulse-Shapers for Single-Channel QPSK

The performance of digitally, electrically, and optically pulse-shaped signals is investigated first. As a signal source for all of the pulse-shaping techniques we use a single-polarization QPSK signal encoded onto a single-carrier. The initial pulses have a non-return-to-zero (NRZ, rectangular) pulse shape. NRZ-QPSK signals have been chosen because they are widely employed and show good performance especially for long haul transmission. The experimental setup is depicted in Fig. 5.1. To generate the NRZ-QPSK data pulses we use a versatile software-defined optical Tx comprising two Xilinx XC5VFX200T field programmable gate arrays (FPGA) and two high-speed Micram digital-to-analog converters (DAC). The DACs are operated at sampling rates up to 30 GSa/s with a physical resolution of 6 bit and an analog electrical bandwidth $f_{\mathrm{DAC}} > 18$ GHz. The respective pulse-shaping for the three schemes is implemented as follows:

- The digital filters (marked green in Fig. 5.1) are realized in the FPGA. The additional electrical filters (red) are then used to remove the digitally generated image spectra (aliasing) when sinc-pulses are generated in the digital domain.
- When the sinc-shape is approximated in the electrical domain the electrical filters alone shape the electrical drive-signals that are fed to the IQ-modulator. The IQ-modulator then encodes QPSK data onto an external cavity laser (ECL, wavelength λ_1, linewidth 100 kHz).

- When performing pulse-shaping in the optical domain the DAC output signals are directly fed to a nested $LiNbO_3$ Mach-Zehnder IQ-modulator (MZM) with a modulation bandwidth of $f_{MZM} > 25$ GHz. To generate the optically filtered QPSK signals, a Finisar WaveShaper serves as pulse-shaper located behind the modulator (marked blue in Fig. 5.1).

A variable optical attenuator (VOA) adjusts the optical power launched into the first erbium doped fiber amplifier (EDFA), and thus varies the optical signal-to-noise ratio (OSNR in a bandwidth of 0.1 nm).

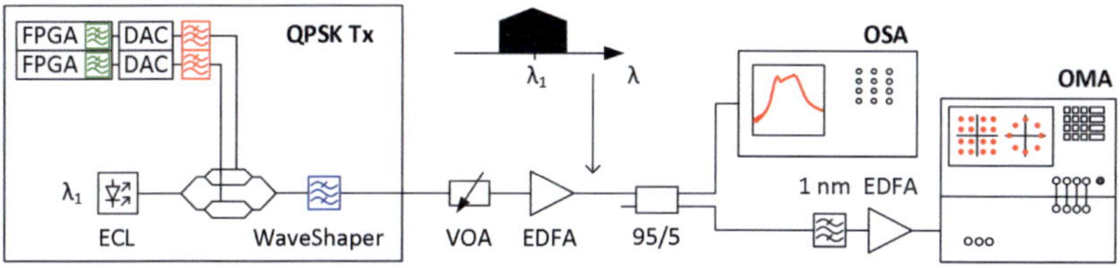

Fig. 5.1 Setup for single carrier pulse-shaping and measurement of its performance. A pair of field programmable gate arrays (FPGA) drives two high-speed digital-to-analog converters (DAC). The digital pulse-shaper (green) is realized within the FPGAs. Electrical anti-alias filters (red) remove either alias spectra for the digital pulse-shaper or solely perform pulse-shaping of the QPSK signals. The electrical signals then modulate an external cavity laser (ECL) by means of an optical I/Q-modulator. In the case of optical pulse-shaping a Finisar WaveShaper (blue) is employed. We use a variable optical attenuator (VOA) together with an erbium doped fiber amplifier (EDFA) to adjust the optical signal-to-noise ratio (OSNR). A 95/5 splitter directs the signals to an optical spectrum analyzer (OSA) and to a coherent receiver (OMA, Agilent optical modulation analyzer).

A schematic optical power spectrum centered at the ECL wavelength λ_1 is shown as an inset. The spectrum drops toward the band edges. In the case of digital filtering, this is due to the frequency response of the DAC and the anti-alias filters (an influence which could have been compensated for by digital pre-conditioning). For the electrical and optical filters this spectral drop cannot be avoided in practice, and the spectral cut-off cannot be as sharp as for the digital filter. However, this non-ideal spectral shape can be compensated in the Rx as will be explained in the next subsection.

An optical spectrum analyzer (OSA) determines the OSNR. The signal is filtered by a standard 1 nm optical filter which removes spurious EDFA noise. Finally, the signal power is leveled with the second EDFA and coherently received by the Agilent optical modulation analyzer (OMA). The OMA comprises two 90° optical hybrids (one for each polarization) and balanced photo-detectors. A free-running ECL serves as an internal local oscillator (LO). The signals are sampled by real-time oscilloscopes with 80 GSa/s each having an analog bandwidth of 32 GHz.

Digital Filtering in the Transmitter: Digital Nyquist pulse-shaping has proven excellent performance in ultra-densely spaced WDM networks. The digitally sinc-shaped Nyquist pulses have been generated by our software defined Tx which acts as an arbitrary waveform generator (AWG), i.e., signal generation and digital pulse-shaping (Fig. 5.1, green) is performed offline. An FIR filter of order $R = 2048$ was used for pulse-shaping. The generated signals are then stored in the FPGAs. A 6 bit DAC provides the transition from the digital to the analog domain. The electrical anti-alias low-pass filters of $f_{\mathrm{el}} \approx 12.3$ GHz (Fig. 5.1, red) remove spurious image spectra created by the DACs, and we end up with sinc-shaped Nyquist pulses with virtually zero spectral roll-off. The DACs are operated at 30 GSa/s and the symbol rate is 20 GBd leading to an effective oversampling factor of $q = 30 / 20 = 1.5$.

Ensemble averaged spectra measured with the OMA are shown in Fig. 5.2(a). Due to the high filter order R, nearly all of the signal power is confined to the Nyquist frequency band (Fig. 5.2(a), top). Thus the optical signal bandwidth is virtually 20 GHz for a 20 GBd QPSK signal. Static and dynamic equalization in the Rx flattens the spectral roll-off. The resulting spectrum shows a flat Nyquist pass-band and steep edges (Fig. 5.2(a), bottom).

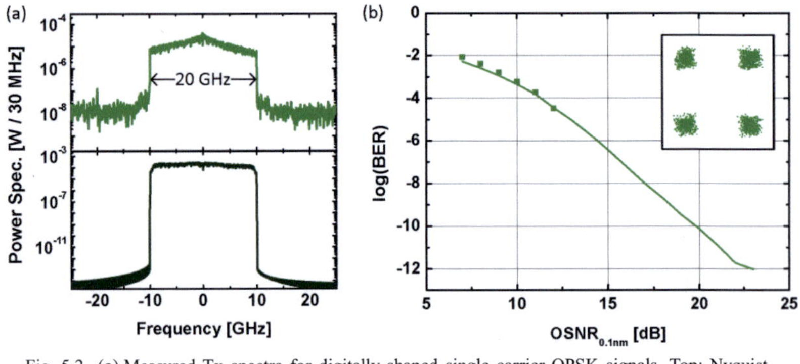

Fig. 5.2 (**a**) Measured Tx spectra for digitally shaped single-carrier QPSK signals. Top: Nyquist-shaped Tx signal spectrum, not compensated for DAC roll-off. Bottom: Rx signal spectrum after Nyquist filtering and additional static and dynamic equalization which compensate for the roll-off introduced by Tx and Rx electronics. (**b**) Measured BER performance for digitally shaped single-carrier QPSK signals. Measured BER (squares) and estimated BER as derived from EVM measurements (solid line) as a function of OSNR for 20 GBd QPSK. The inset shows the constellation diagram at the highest possible OSNR of 30 dB.

We measure BER and EVM as a function of OSNR. We further estimate an equivalent BER from the measured EVM and display the results in Fig. 5.2(b). Measured BER (squares) and estimated BER derived from EVM (line) coincide. An inset shows a constellation diagram for the highest achievable OSNR of 30 dB.

Electrical Filtering: To generate sinc-shaped Nyquist pulses in the electrical domain we use (as approximation to rectangularly shaped filters) the same low-pass anti-alias filters as before. Although we keep the DACs in the setup, they only produce two-level NRZ electrical signals. Therefore binary drivers suffice, which potentially reduces overall cost of the Tx significantly. The simulated S-parameters provided by the manufacturer and the group delay derived from the transfer function S_{21} of the electrical low-pass filters are depicted in Fig. 5.3(a) and (b). A 3 dB cut-off frequency of f_{el} = 12 GHz can be seen from the 20 log10$|S_{21}|$ curve (blue). The reflection represented by the 20 log10$|S_{11}|$ curve (red) is suppressed by at least 20 dB throughout the pass-band, Fig. 5.3(a). The actually manufactured filters show 3 dB cut-off frequencies of $f_{el} \approx$ 12.3 GHz. The group delay in the pass-band stays below 0.5 ns, see Fig. 5.3(b).

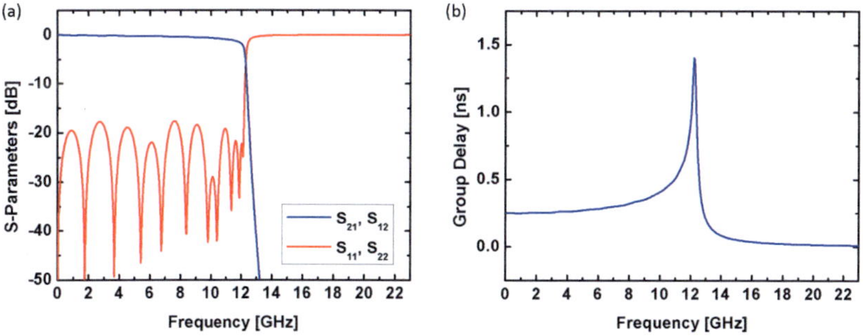

Fig. 5.3 (**a**) Simulated S-parameters of the employed electrical low-pass filters. The measured cut-off frequency of the manufactured filters is $f_{el} \approx$ 12.3 GHz. (**b**) Simulated group delay of the employed electrical low-pass filters. The group delay stays below 0.5 ns throughout the pass-band.

A typically received and ensemble averaged power spectrum of a 20 GBd QPSK signal at an OSNR of 30 dB is shown in Fig. 5.4(a), upper row. In order to obtain the spectrum, we measure the time domain waveform with the OMA and perform a Fourier transform. The optical signal bandwidth is $f_{opt} \approx$ 25.2 GHz close to the 2 × $f_{el} \approx$ 24.6 GHz pass-band of the electrical filters. After filtering and performing the Rx DSP the signal bandwidth is digitally reduced to the Nyquist frequency band of 20 GHz, and the pass-band is flattened as to be seen in Fig. 5.4(a), lower row. Constellation diagrams for 20 GBd and 24 GBd QPSK are shown in Fig. 5.4(c).

Fig. 5.4(b) shows the measured BER (squares) and the BER estimated from EVM (solid lines). Measured and estimated BER agree well. Measurements are done for single-polarization and single-carrier QPSK, for different symbol rates, and for different OSNR. As expected, the BER degrades with increasing symbol rate. For large OSNR, a BER error floor can be seen. This error floor stems from the electronic noise originating from the Tx and Rx. However, this noise is negligible compared to the optical noise of multiple EDFAs that will be picked in transmission links with multiple amplifiers. For the given optical output power of our Tx with electrical pulse-shaping, and for the given EDFA, the maximum achievable OSNR is 34 dB.

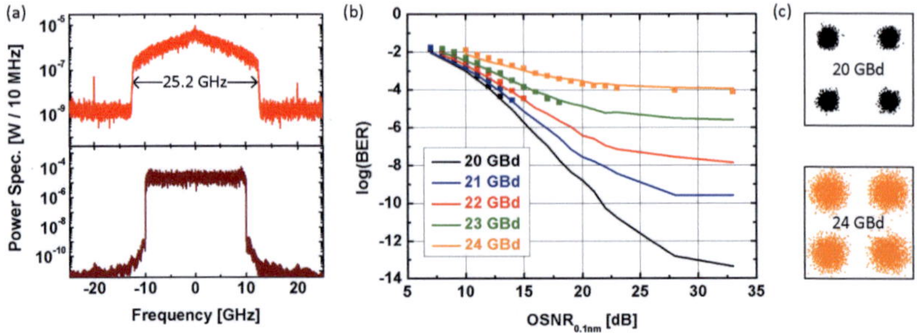

Fig. 5.4 Measured and ensemble averaged spectrum (top) of a 20 GBd electrically generated QPSK signal. As expected, the overall signal bandwidth corresponds to two times the electrical filters' cut-off frequency. Measured spectrum (bottom) for the same signal after applying the Nyquist filtering procedure in the Rx as described in Fig. 2. The filter removes signal components outside the Nyquist frequency band and flattens the spectrum in the pass-band. (**b**) Measurement results for electrically shaped QPSK signals at varying symbol rates as a function of OSNR. Measured BER (squares) and estimated BER(EVM) (solid lines) for different symbol rates and different OSNR. The error floor at high OSNR values stems from the electrical noise added by Tx and Rx. This noise has negligible influence on the measurements. (**c**) Received constellation diagrams for 20 GBd (top) and 24 GBd (bottom) QPSK.

The overall BER as a function of OSNR is approximately same comparing a single optical carrier modulated with either digitally or electrically shaped sinc-pulses. However, the digital pulse-shaper produces 20 GBd signals with virtually 20 GHz bandwidth whereas electrically shaped 20 GBd signals require a bandwidth of 25.2 GHz. As a disadvantage, the maximum achievable OSNR for the digital pulse-shaper is 4 dB less than for the analog pulse-shaper. This is dominantly due to the increased peak-to-average power ratio of the digitally shaped signals with their more pronounced side lobes as compared to the pulses shaped by analog filters. For the multi-carrier experiments in the next section we limit the symbol rate to 20 GBd, since already a symbol rate of 21 GBd leads to a significant penalty.

Optical Filtering: Finally, to approximate sinc-shaped pulses in the optical domain we use a Finisar WaveShaper as an optical band-pass filter and apply it to conventional NRZ-QPSK signals. We adjust the WaveShaper to have a fixed optical pass-band of 12.5 GHz. Since the filter pass-band at this resolution is difficult to change we instead vary the symbol rate from 20 GBd to 28 GBd in steps of 2 GBd. We only measure signals where the signal quality remains above the quality required for state-of-the-art forward error correction (FEC).

The spectrum for a 20 GBd optically filtered QPSK signal can be seen in Fig. 9(a), upper row. As expected and due to the Lorentzian shape of the optical filter we do not see steep band edges as for the electrically shaped signal spectrum in Fig. 5.4(a). After DSP at the Rx we again obtain a flat pass-band of the signal spectrum (Fig. 5.5(a), lower row) leading to a minimum ISI. Constellation diagrams for 20 GBd and 24 GBd QPSK are shown in Fig. 5.5(c). Measured BER and estimated BER(EVM) for different OSNR and different symbol rates are shown in Fig. 5.5(b). The BER increases with increasing symbol rate as the fixed filter width of 12.5 GHz significantly affects signals faster than 22 GBd. Applying the Nyquist filtering at the Rx as described in Section II.A decreases BER and EVM for signals with 28 GBd. For smaller symbol rates we find that there is no difference for a Rx with and without said electronic Nyquist filtering technique. Since 20 GBd signals showed best performance we use these signals for the WDM experiments described in Section III. The highest achievable OSNR decreases from 30 dB to 25 dB which reflects the additional insertion loss of the WaveShaper in our setup.

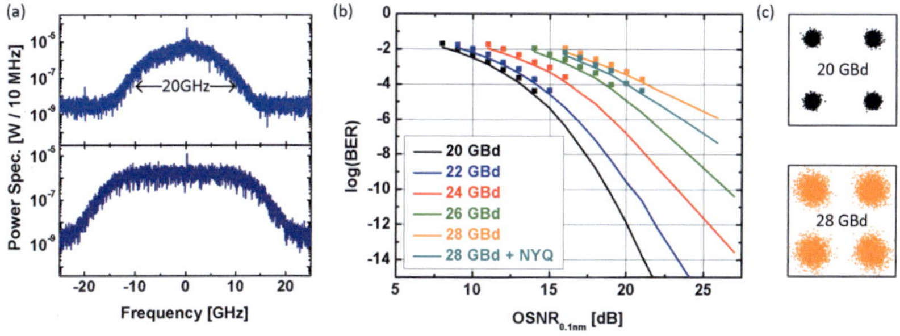

Fig. 5.5 (**a**) Measured and ensemble averaged spectrum (top) of a 20 GBd optically shaped QPSK signal. The spectral roll-off is not as steep as for electrically shaped signals. After DSP equalization at the Rx, the pass-band of the signal is flat (bottom). (**b**) Measurement results for optically shaped QPSK signals at various symbol rates as a function of OSNR. Measured BER (squares) and estimated BER(EVM) (solid lines) for different symbol rates and different OSNR. In our optical filter setup Nyquist filtering at the Rx only provides an advantage for symbol rates ≥ 28 GBd. (**c**) Received constellation diagrams for 20 GBd (top) and 28 GBd (bottom) QPSK.

Comparing Pulse-Shapes: In order to give a better idea of how accurately a sinc-shaped impulse form is met when employing digital, electrical, and optical pulse-shapers, we use the received signal spectra from Fig. 5.2–5.5 (upper rows) without any equalization and derive the individual pulse forms. The outcome is depicted in Fig. 5.6. The digitally shaped pulse in Fig. 5.6 (top row) most accurately approximates a sinc-shaped impulse. The electrical pulse-shaper still produces sinc-typical side lobes but they decay rapidly, see Fig. 5.6 (middle row). The optical pulse-shaper yields the worst sinc-approximation, see Fig. 5.6 (bottom row). This was expected as the transfer of the optical filter is Lorentzian and not rectangular. For high symbol rates (≥ 50 GBd), where several segments of the WaveShaper are transparent, the overall filter approximates a rectangle much better than for a 12.5 GHz single segment pass-band.

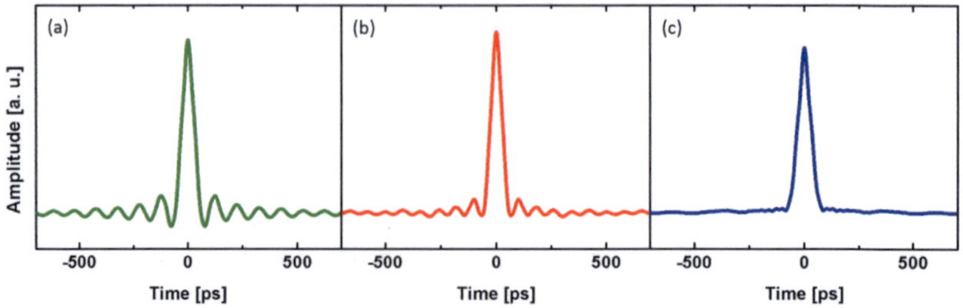

Fig. 5.6 Different pulse forms measured from digital (top, green), electrical (middle, red), and optical (bottom, blue) pulse-shapers. As expected, the digital pulse-shaper approximates a sinc-shaped pulse form most accurately. The electrical pulse-shaper still produces sinc-typical side lobes, whereas the optical pulse-shaper matches a sinc-function worst.

5.1.1.2 Performance Evaluation of the Different Pulse-Shapers in a WDM Network

The differently shaped QPSK signals are now employed in an ultra-dense WDM network scenario emulated by three channels with different, free-running carrier frequencies. A common QPSK symbol rate of 20 GBd is chosen for each channel. The channel spacing was varied between 17 GHz and 50 GHz and thus covers the WDM as well as the Nyquist WDM case. We measure single-polarization and PDM signals. The quality of filtered, band-limited Tx signals is compared to the standard rectangular NRZ pulses, which are either received as is, or rectangularly filtered at the Rx.

For evaluating the performance of the different pulse-shaping techniques we investigate the transmitters discussed in the previous section within a three-carrier ultra-dense WDM setup, Fig. 5.7.

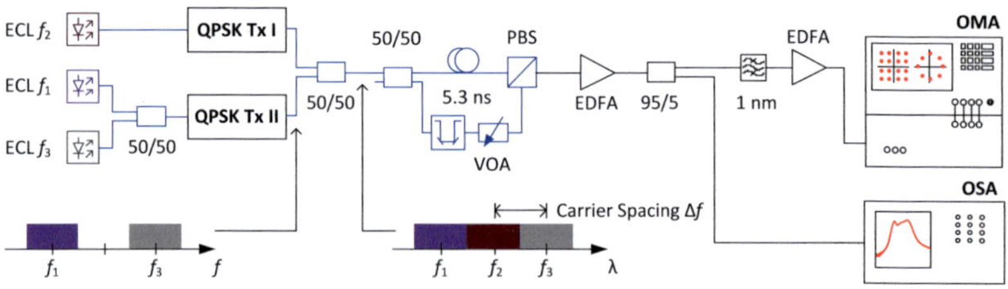

Fig. 5.7 Setup to test the minimum required carrier spacing $\Delta\lambda$ in an ultra-dense WDM network. Three free-running ECLs are encoded with QPSK signals and shaped in the digital, the electrical and optical domain. For testing purposes an unshaped NRZ-QPSK is tested as well. Two independent Tx guarantee uncorrelated data in adjacent channels. A worst-case scenario with polarization maintaining components (blue optical paths) maximizes inter-channel crosstalk. For PDM experiments the combined three channels are split, delayed by 5.3 ns, and combined in orthogonal polarizations. The remaining setup is identical to the ones used in the previous section.

Three external cavity lasers (ECL) provide the three optical carriers. A fourth ECL is used as LO within the OMA. All lasers are free-running, i.e., there is neither frequency nor phase locking. Two different transmitters guarantee de-correlated data streams for the middle channel (Tx I) and the two outer channels (Tx II). The three signals are combined with equal powers. Polarization division multiplexing is emulated by splitting the signals in two arms, applying a delay of 5.3 ns in one arm, and finally combining both arms two form two orthogonal polarizations. For a worst-case linear cross-talk (where adjacent channels have the same state of polarization) we use polarization maintaining components and fibers (Fig. 5.7, blue). The ultra-dense WDM signal is amplified and coherently received by the OMA. By varying the carrier spacing $\Delta\lambda$ we determine the potential of the three pulse-shaping techniques for the minimum guard band and thus best spectral efficiency (SE) in a WDM network. The evaluation is based on both BER and EVM. Measured spectra for pulses shaped with different techniques at a channel spacing of $\Delta\lambda$ = 25 GHz and for a symbol rate of 20 GBd are shown in Fig. 5.8. As a reference we first depict the unfiltered NRZ signal, Fig. 5.8(a). The NRZ signal is only shaped by the limited electrical bandwidths of DACs (> 18 GHz) and the bandwidth of the optical modulator (25 GHz). For digitally pulse-shaped QPSK signals, see Fig. 5.8(b), the filter slopes are so steep that even notches appear in the region between the channels. For the case of the QPSK signals shaped electrically, see Fig. 5.8(c), and optically, see Fig. 5.8(d), one can see that the three channels slightly overlap.

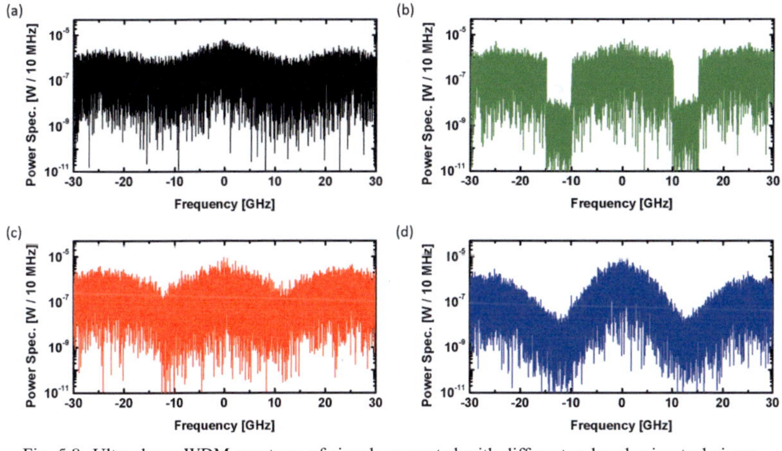

Fig. 5.8 Ultra-dense WDM spectrum of signals generated with different pulse-shaping techniques. The carrier spacing is $\Delta\lambda$ = 25 GHz and the carriers are QPSK encoded with single-polarization 20 GBd. (**a**) Unfiltered NRZ pulse-shape, only shaped by the limited electrical bandwidth of the DACs (> 18 GHz) and the optical modulator (25 GHz). (**b**) Spectrum of digitally generated sinc-shaped QPSK signal. The digitally generated signal shows distinct spectral notches which are due to the steep-edged digital filters in the Tx. (**c**) Electrically pulse-shaped QPSK spectrum. (**d**) Optically pulse-shaped QPSK spectrum.

We determine BER and BER derived from EVM measurements for all pulse-shaping schemes and for varying channel spacing Δλ. All signals have a symbol rate of 20 GBd and are transmitted with highest possible OSNR. The results for the single polarization and the dual polarization experiments are depicted in Fig. 5.9(a) and (b), respectively. We begin with the unfiltered, plain NRZ signal (black solid lines for EVM derived BER and squares for BER measurements). As expected, the impact of inter-channel interference (ICI) for unfiltered, plain NRZ is largest. We also applied the Nyquist filtering technique to the unfiltered NRZ signal at the Rx (brown). Both signals require larger channel spacing than any of the pulse-shaped signals. At Δλ = 30 GHz we see a break-even point for the unshaped NRZ signals. There is a local maximum of the Rx-filtering curve (brown) near Δλ = 36 GHz where the unfiltered sinc-shaped Tx spectra start overlapping. It seems that there is a difference between detection with or without Rx Nyquist-filtering. This local maximum is much more distinct for the Rx Nyquist-filtered signals (brown) but also visible for the conventionally received NRZ (black). Optically pulse-shaped (blue) and electrically pulse-shaped (red) signals show negligible ICI for Δλ ≥ 25 GHz. We attribute the increased error floor for optically filtered PDM-QPSK (Fig. 5.9(b), blue) to the polarization de-multiplexing algorithm, which is sensitive to signal components of the two outer channels within the received middle channel. This is especially critical for optically pulse-shaped signals where the filter slopes are not very steep. The digitally filtered signals show negligible penalty due to ICI up to the Nyquist channel spacing of Δλ =20 GHz. This technique is clearly best suited for Nyquist WDM setups where the channel spacing is equal to the symbol rate and the symbol rate is sufficiently low. However, this digital operation comes at a price of intense signal processing so that electrical and optical pulse-shaping most likely have a CAPEX and OPEX advantage.

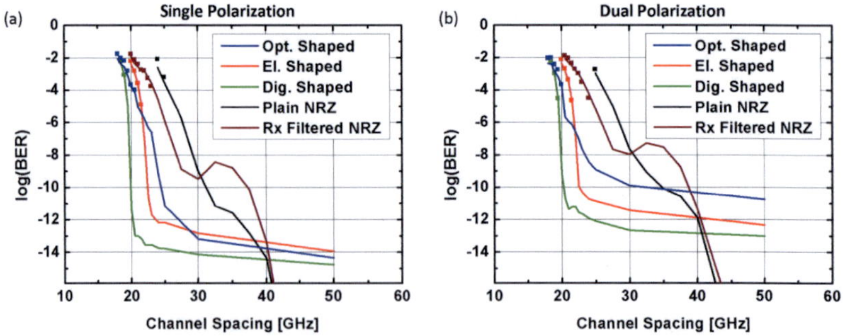

Fig. 5.9 Measured BER (squares) and estimated BER(EVM) (lines) for different pulse-shaping techniques as a function of varying channel spacing Δλ. All channels transmit 20 GBd QPSK signals. The local maximum for Nyquist Rx-filtering near Δλ = 36 GHz marks the point where the unfiltered sinc-shaped Tx spectra start overlapping. (**a**) Single polarization setup. (**b**) Polarization division multiplexing (PDM).

5.1.1.3 Conclusion

We investigated the performance of digitally, electrically, and optically pulse-shaped QPSK signals for single-carrier transmission and in a three-carrier ultra-dense WDM set-up. For this purpose, both BER and EVM were measured. Digitally shaped sinc-pulses outperform other pulse-shaping techniques that rely on current state-of-the art electrical or optical filters. For digitally shaped signals the crosstalk is negligible even for a channel spacing of $\Delta\lambda = 20$ GHz corresponding to the Nyquist limit for 20 GBd signals. Yet, it is important to note that both electrically and optically pulse-shaped signals always outperform unfiltered NRZ in terms of spectral efficiency. As an advantage of the analog techniques compared to digital pulse-shaping, costly DACs are not required and power consumption can be significantly reduced.

5.1.2 512QAM Nyquist Signaling with Record Spectral Efficiency

512QAM Nyquist sinc-pulse transmission at 54 Gbit/s in an optical bandwidth of 3 GHz

R. Schmogrow, D. Hillerkuss, S. Wolf, B. Bäuerle, M. Winter, P. Kleinow, B. Nebendahl, T. Dippon, P. Schindler, C. Koos, W. Freude, and J. Leuthold

Optics Express 20(6), pp. 6439–6447 (2012). [J17]

Sinc-shaped Nyquist pulses [147][79] are attractive to achieve high spectral efficiencies. They overlap in time but follow each other at such a symbol period T_s so that they do not interfere according to the Nyquist inter-symbol interference (ISI) criterion. With this condition met, no inter-symbol crosstalk occurs. The resulting signal spectrum has an ideal rectangular shape. Therefore, sinc-shaped signals require the minimum Nyquist bandwidth [61], a fact which is beneficial in networks with multi-carrier or wavelength division multiplexing (WDM) as is common today [148][149][150]. The rectangular spectral shape allows decreasing the WDM channel frequency spacing up to the point where the spacing equals the symbol rate. This scenario is known as Nyquist WDM [24][31]. Sinc-shaped Nyquist signals are also often described by a so-called "zero roll-off" characteristic of raised cosine-shaped spectra [151].

Nyquist pulse shaping has much in common with orthogonal frequency division multiplexing (OFDM) [23][39]. In fact, both are related by a duality principle, so that time and frequency can be simply interchanged [79]. Both Nyquist WDM and OFDM converge towards the same high spectral efficiency (SE) for large OFDM subcarrier and WDM tributary counts, but only Nyquist pulse shaping offers SE advantages for low tributary numbers [79]. Furthermore, Nyquist pulse shaping has some additional advantages: In contrast to OFDM, where subcarriers overlap and thus must be transmitted and processed in synchronism, Nyquist WDM tributaries can be transmitted and processed asynchronously [31]. Further, calculations indicate [79] that Nyquist pulse shaping leads to lower peak-to-average power ratios (PAPR) than OFDM. This could prove advantageous in transmission if nonlinearities play a role, and on reception with a view to the required resolution of the analog-to-digital converters (ADC). Signals having a high PAPR naturally need to be quantized with a high resolution in order to precisely account for large and small amplitudes at the same time.

The spectral efficiency of the transmission can be further increased by using polarization shift keying or polarization division multiplexing (PDM) [78] along with advanced modulation formats such as *M*-ary quadrature amplitude modulation (QAM). Recently, QAM signals in combination with Nyquist pulse shaping using a roll-off factor of $\beta = 0.35$ have been demonstrated. With this technique 54 Gbit/s were transmitted in 512QAM format inside an optical bandwidth of only 4.1 GHz [61]. Appropriate so-called raised-cosine filters are described in [74], p. 858, and the resulting signal bandwidth B in dependence on the symbol duration T_s (symbol rate $1 / T_s$) and the roll-off factor β follows from

$$B = \frac{1}{T_s}(1+\beta). \tag{3.1.1}$$

In this paper we demonstrate a record transmission of 54 Gbit/s within an optical bandwidth of only 3 GHz corresponding to a symbol rate of $1/T_s = 3$ GBd. This is achieved by employing Nyquist sinc-pulse shaping and a PDM-512QAM format. Our sinc-shaped pulses are basically generated with a roll-off factor of $\beta = 0$, but technically they are limited to a finite time window of 512 symbol durations T_s. Taking that into account, we find from simulations and from [79], Eq. 6 an actual roll-off $\beta = 0.0024$. The technique enables Transmission distances of up to 44 km which were reached by employing ultra-large area fiber (ULAF).

We further demonstrate the transmission of 48 Gbit/s in a 3 GHz optical channel over a distance of 150 km in a standard single mode fiber (SSMF) using the Nyquist shaped PDM-256QAM modulation format.

5.1.2.1 Experimental setup

The theoretical zero roll-off sinc-shaped Nyquist signal is schematically shown in Fig. 5.10(a) (green). A single impulse together with its corresponding spectrum is displayed. The measured, ensemble averaged spectrum of a 3 GBd Nyquist sinc-shaped ($\beta = 0.0024$) signal occupies only a bandwidth of virtually 3 GHz, see Fig. 5.10(b) and Eq. (3.1.1). The power spectrum has been measured with a coherent receiver. The noise floor is due to quantization noise at the transmitter and the receiver, which employed digital-to-analog and analog-to-digital converters with a nominal resolution of 12 bit and 8 bit, respectively. We see an impressive out-of-band suppression of more than 36dB. Therefore linear inter-channel crosstalk between such signals is negligible in WDM networks [31][152].

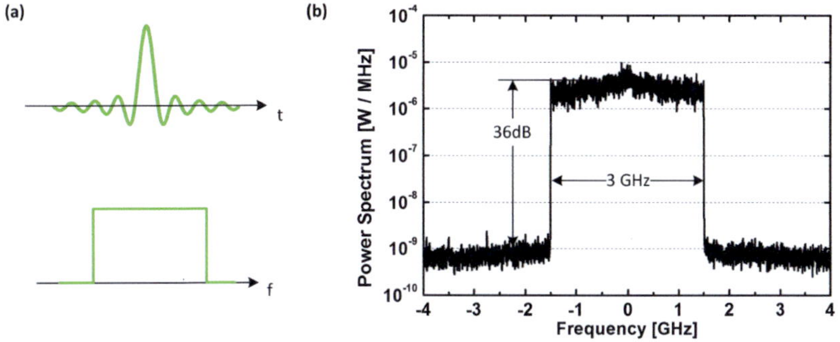

Fig. 5.10 Schematic of elementary impulse and corresponding spectrum along with signal spectrum in one polarization only. (**a**) sinc-shaped Nyquist impulse and rectangular spectrum. (**b**) Power spectrum of sinc-shaped Nyquist-pulses belonging to a random sequence of constellation points, measured with a coherent receiver and in one polarization. The 14 dB signal bandwidth is 3 GHz for a symbol rate of 3 GBd. The noise floor is due to signal quantization. The out-of-band suppression is more than 36dB.

The experimental Nyquist transmission setup is depicted in Fig. 5.11. The Nyquist transmitter (Tx) comprises an arbitrary waveform generator (AWG), a narrow linewidth (< 1 kHz) laser source and an optical I/Q-Modulator, see Fig. 5.11(a). The sinc-shaped Nyquist pulses are generated off-line and stored in an AWG (Agilent M8190A) which has a nominal resolution of 12 bit at a sampling rate of 12 GSa/s, and an electrical bandwidth of 5 GHz. Pseudo random binary sequences (PRBS) show a periodicity of 2^N-1 whereas memory based signal generators such as AWGs support even-numbered length patterns only. To overcome this issue when employing standard bit error ratio testers (BERT), a rather short PRBS is replicated until the periodicity of concatenated PRBS patterns agrees with the periodicity required by the memory. For high-order modulation formats such as 256QAM or 512QAM, very long PRBS patterns (e.g. $2^{31}-1$) would be needed in order to assure that all points in the constellation diagram are hit appropriately often. This would require an excessively large AWG memory. Therefore, in this experiment we use random data generated externally by Matlab. We generate new sets of random data for each measurement thereby approximating real data traffic. The generated signals are four-times oversampled to facilitate the removal of periodic spectral repetitions by anti-alias filters. Alternatively, passive optical filters [121] or digital filters with lower oversampling factors can be used to remove spurious spectral tributaries outside of the Nyquist frequency band [19]. The two outputs of the AWG directly drive the in-phase (I) and quadrature (Q) inputs of the optical I/Q-modulator. Nonlinearities introduced by the modulator's sinusoidal transfer function do not significantly broaden the signal spectrum, because the modulator is differentially driven with 700 mV only ($V_\pi \approx 4$ V), and is therefore operated well in its linear regime. A fiber laser (laser 1) with approximately 1 kHz linewidth (nominal wavelength: 1550 nm) serves as a source for the Tx. The inset of Fig. 5.11(a) shows a schematic eye-diagram of a two-level zero roll-off Nyquist signal. It can be seen that the amplitude of the field is not constant over the symbol duration T_s so that the information can be recovered without error at the point of the maximum eye opening. The tolerance of zero roll-off Nyquist signals with respect to jitter and sampling phase error as well as a novel clock phase recovery algorithm have been discussed in [79].

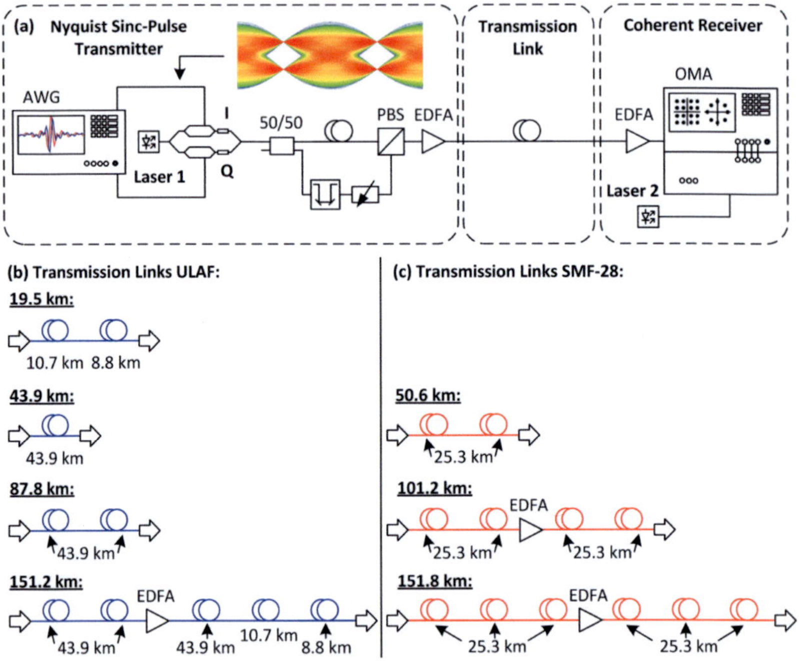

Fig. 5.11 Transmission setup. (**a**) An optical transmitter based on an Agilent arbitrary waveform generator (AWG) generates, with the help of a narrow-linewidth fiber laser (laser 1), a variety of Nyquist sinc-pulses. The inset shows a schematic eye-diagram for a two-level zero roll-off Nyquist signal. The optical signal is split, de-correlated, and recombined in orthogonal polarizations to emulate polarization division multiplexing (PDM). Several fiber spans are inserted in-between two EDFAs before the signal is received by an Agilent optical modulation analyzer (OMA) using a second narrow-linewidth fiber laser (laser 2). (**b**) Several ULAF based transmission links that were used for our experiments. (**c**) Several SSMF based transmission links that were used for our experiments.

A PDM signal is generated within the next stage. The optical signal leaving the I/Q-modulator is split onto two arms, delayed by 16 ns in order to de-correlate the signals, and then combined in orthogonal polarizations. Due to the zero roll-off of the signals, an optical bandwidth of only 3 GHz at a symbol rate of 3 GBd is required as seen in Fig. 5.10(b). As a consequence, virtually all of the signal power is confined to the Nyquist frequency band $B = 1 / T_s$.

During the experiment, we increase the transmission link length in steps up to the full length of 150 km. SMF-28 and OFS ULAF fibers are used. All combinations of SMF-28 and ULAF together with the locations of the EDFAs are shown in Fig. 5.11(b) and (c). At 1550 nm the SMF-28 has a typical loss of 0.19 dB/km with a dispersion of 16.8 ps/nm/km, whereas the ULAF typically attenuates by 0.18 dB/km and comes with a dispersion of 20.2 ps/nm/km. The effective area of the ULAF is specified to be approximately 130 μm^2.

In the receiver (Rx) we use an EDFA to pre-amplify the signals before they are fed into a polarization-diversity intradyne system, the input power to which was kept constant throughout all measurements. A second fiber laser (laser 2) with a linewidth of approximately 200 Hz (centered at 1550 nm) serves as a local oscillator (LO). An Agilent N4391A optical modulation analyzer (OMA) is used to record the incoming signals. We slightly detune the bias of the modulator in the Tx and allow a residual carrier which serves as reference for frequency offset compensation. No additional pilot tone was sent over the link.

All digital signal processing (DSP) and signal evaluation was performed offline. The processing included chromatic dispersion compensation [153][154], separation of the two polarizations [71], channel equalization, and a new special zero roll-off clock phase recovery [79]. The sampled complex signals of X- and Y-polarization are equalized by an FIR filter which removes accumulated dispersion associated with the properties of used fiber spans. As a next step the signal is analyzed in the Stokes parameter space and aligned to the polarization axis of the receiver [71]. This algorithm does not compensate for polarization mode dispersion, and no feedback regarding the signal quality is given. The clock phase recovery assures that the signal is sampled at the point of maximum eye opening. Carrier phase recovery is performed by compensating a residual frequency offset of the signal, corresponding to a linear phase over time, in the frequency domain. Then the constellation diagram is aligned to the real- and imaginary axis of the OMA by minimizing the error vector magnitude (EVM). Finally, a linear FIR equalizer with 55 taps is trained through a feedback mechanism and optimizes the signal quality by minimizing the EVM. This equalizer removes pattern effects as well as it compensates for the linear impulse response of the used components (DAC, modulator, ADC, and amplifiers). Nonlinear algorithms (e.g. digital back propagation) have not been used.

5.1.2.2 Results

Performance compared to the Shannon limit: The limiting capacity C for error-less transmission over a linear channel with Gaussian distributed signal and additive white Gaussian noise (AWGN) is given by [155]

$$C = B\log_2\left(1+\text{SNR}\right). \tag{3.1.2}$$

The quantity B denotes the channel bandwidth (here: $B = 3$ GHz, see Fig. 5.10(b)), and SNR is the signal-to-noise power ratio for Gaussian distributed signal and noise. For an SNR of 37 dB as measured with the coherent receiver for one polarization only, this limiting channel capacity amounts to $C = 37$ Gbit/s. The maximum spectral efficiency SE_{max} per polarization is obtained by relating the limiting channel capacity Eq. (3.1.2) to the channel bandwidth,

$$\text{SE}_{\text{max}} = C/B = \log_2\left(1+\text{SNR}\right). \tag{3.1.3}$$

With the data above, a maximum SE_{max} of 12.3 bit/s/Hz results per polarization. We actually measured 0.5×56 Gbit/s/3 GHz = 9 bit/s/Hz per polarization, i.e., 18 bit/s/Hz for PDM-512QAM data in a hypothetical WDM network, at a worst-case BER of 1.9×10^{-2}. To compare to the Shannon limit, we have to employ a forward-error correction (FEC) scheme with either 6.7 % overhead for a BER up to 3×10^{-3}, yielding an effective BER of 10^{-16} [156], or a 25 % overhead for BER up to 1.9×10^{-2} resulting in an effective BER of 1.1×10^{-15} [132]. The resulting BER are small enough to approximate an error-free transmission.

Consequently, our effective SE reduces from a raw value of 18 bit/s/Hz to $\text{SE}_{1.9\times10^{-2}} = (18 / 1.25) = 14.4$ bit/s/Hz and $\text{SE}_{3\times10^{-3}} = (16 / 1.067) = 15$ bit/s/Hz, respectively. After transmission of these signals over 44 km of ULAF, a SNR = 36 dB is recorded. For two polarizations the Shannon limit then would be $2\times\text{SE}_{\text{max}} = 2\times12$ bit/s/Hz according to Eq. (3.1.3), still significantly larger than our optimum $\text{SE}_{3\times10^{-3}} = 15$ bit/s/Hz. Effectively, we are limited by receiver noise as the input power to the receiver was lower for transmission than in the back-to-back case.

Signal quality evaluation for back-to-back and transmission experiments: Measurements were performed for a transmission up to 150 km for 256QAM and up to 44 km for 512QAM using an ULAF and a SSMF fiber, respectively.

The first row of Fig. 5.12 shows error vector magnitude (EVM) plots as a function of the different transmission link lengths as depicted in Fig. 5.11(b) and (c). For a better legibility, dashed horizontal lines indicate bit error ratios (BER) calculated from the measured EVM [81][82] for common forward error correction (FEC) limits, see Fig. 5.12(a) and (b). Since we use random input data, a non data-aided reception leading to the EVM metric is obligatory. A comment on the reliability in using EVM to estimate BER is to be found in the Appendix. All EVM are plotted for the two polarizations separately (solid and open symbols). Back-to-back measurements are black, SSMF related EVM data are colored red, and ULAF related measurements are colored blue. All received signals show EVM below the FEC limit having a BER equal to 1.9×10^{-2}.

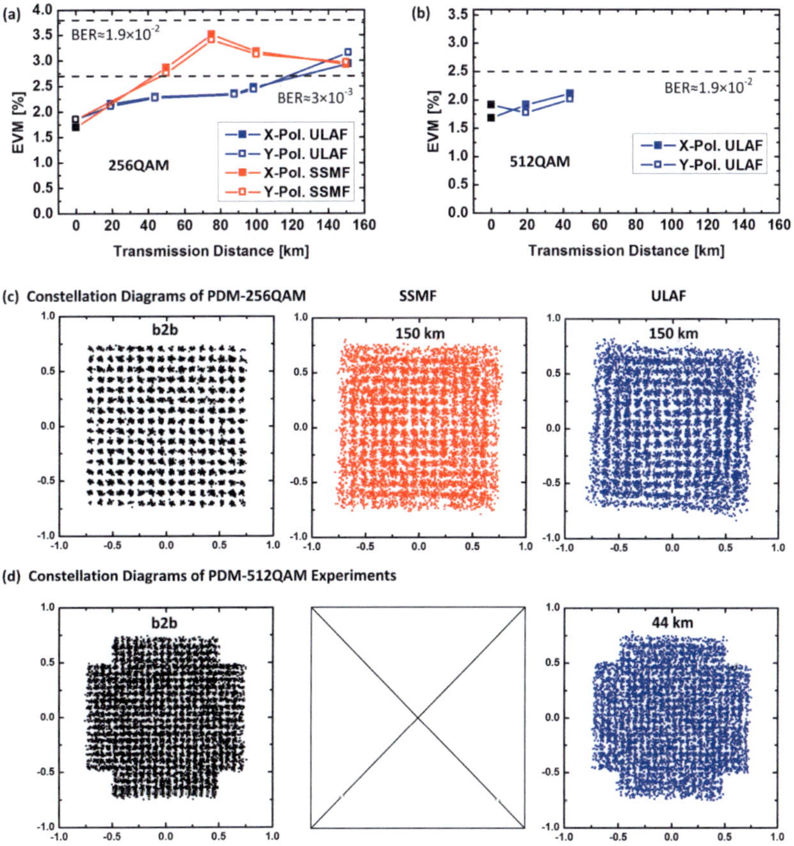

Fig. 5.12 Experimental results for transmitted 256QAM and 512QAM signals. The first row shows the measured EVM of the two polarizations transmitted over a SSMF (red), an ULAF (blue), and back-to-back (b2b) (black). Limits of BER suitable for error-free transmission employing state-of-the-art FEC algorithms are indicated by dashed horizontal lines. (**a**) EVM performance of Nyquist shaped PDM-256QAM data for different fiber spans. (**b**) EVM performance of sinc-shaped PDM-512QAM data for different fiber spans. (**c**) Constellation diagrams for PDM-256QAM and (**d**) PDM-512QAM signals. Both polarizations are plotted within a single constellation diagram. The color coding refers to the type of fiber used for transmission.

The remaining two rows show the aggregate constellation diagrams for both polarizations. Constellations for back-to-back experiments and for transmission over 150 km of either

SSMF or ULAF for the Nyquist shaped PDM-256QAM data are seen in Fig. 5.12(c). The constellation diagram for the 150 km ULAF link shows some non-linear distortions that we attribute to the relatively high launch power as well as imperfect carrier phase synchronization. Therefore, the estimation of the BER by EVM measurements shows some inaccuracy as will be discussed in the next section. The signal performance in the SSMF fiber at 50 km and at 100 km could be improved with further optimization (e.g. by adjusting the launch powers into the fiber spans). The constellation diagrams for sinc-shaped PDM-512QAM Nyquist signals in the back-to-back case and for 44 km ULAF transmission is shown in Fig. 5.12(d). Transmission of the 512QAM format over a SSMF was not possible with a reasonable BER. For all measurements we tested the setup for its tolerance towards polarization rotation and found negligible influence on the signal quality.

Statistical analysis of received data: In order to verify the use of EVM, which requires that signal and noise are Gaussian distributed, we analyze the statistical properties of the received field vectors displayed in Fig. 5.12 with respect to their nominal positions. The deviations from each constellation point are the error vectors, which form a bell-shaped joint probability density function (PDF) as a function of real and imaginary parts of the error vector centered at the respective constellation points. Fig. 5.13(a) shows two neighbored constellation points for the inphase (real part) of the electric field with the respective cross-section PDF in blue and red. We see that for non-data aided reception, as is employed in this work, a decision threshold is required to assign received data to the closest constellation point. This can lead to false assignments and therefore to an underestimation of EVM and BER. However, we already showed that for signals with BER $\leq 10^{-2}$ the EVM difference between non-data aided reception and data aided reception is negligible [81][82].

The statistics of the error vectors for all points in the constellations in Fig. 5.12 were individually checked. The PDF were Gaussian-like, but due to the relative small number of samples (< 100) for each constellation point not very smooth. For better judging how close to a bivariate Gaussian the joint PDF of the error vectors are, we collect all complex error vectors irrespective of the constellation points they were originally associated with. We also disregard the fact that the 61 (72) constellation points of the outermost rows and columns for 256QAM (512QAM) are more noisy than the dominant number of inner ones, which represent a fraction of 76 % (86 %). We relate all error vectors to the field magnitude of the outermost point of the constellation under discussion. We then numerically integrate over the real-part of the error vectors to find the PDF_i of the imaginary part, and repeat this procedure by numerically integrating over the imaginary-part of the error vector to find the PDF_r of the real part. Both, PDF_r and PDF_i, are superimposed and form one histogram, Fig. 5.13(b) and (c).

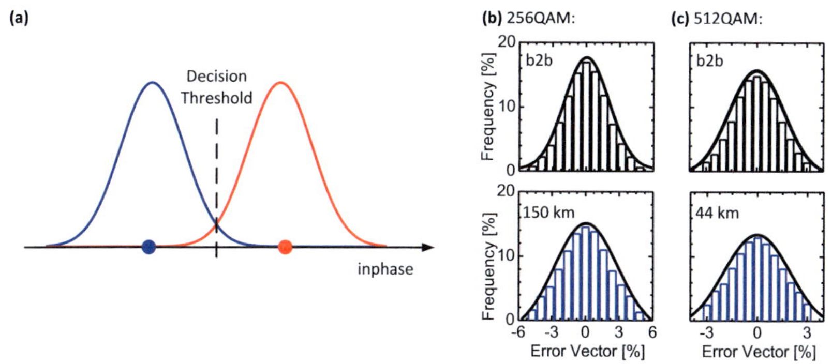

Fig. 5.13 Schematic and measured probability density functions (PDF) approximated by histograms of measured error vectors. (**a**) PDF of two neighboring symbols (blue and red) along the inphase axis for the real part of the electric field, along with their bell-shaped PDF of the error vectors. For non-data aided reception, the required decision threshold could lead to false symbol assignments. For BER $\leq 10^{-2}$ this effect is negligible. (**b**) Histograms approximating a representative PDF both for real and imaginary part of all normalized error vectors of a 256QAM constellation diagram. Upper row: Back-to-back (b2b) transmission, lower row: 150 km ULAF. (**c**) Histograms approximating a representative PDF both for real and imaginary part of all normalized error vectors of a 512QAM constellation diagram. Upper row: Back-to-back (b2b) transmission, lower row: 44 km ULAF.

The frequency of occurrence in Fig. 5.13(b) and (c) is proportional $PDF_r + PDF_i$ as discussed before and resembles a Gaussian closely, which is provided as a solid black fit curve. The respective decision thresholds are located at both the horizontal ends of the error vector axis. It is seen that here the Gaussian is virtually at zero so that interference from neighboring constellation points can be neglected, even for non-data aided reception as discussed in the beginning of this Section. Fig. 5.13(b) and (c) show the results for back-to-back (b2b, upper row) and transmission over an ULAF with 150 km and 44 km length (lower row). These findings support the hypothesis of Gaussian noise dominating the signal quality. For the signal itself the Gaussian hypothesis can be safely assumed based on the central limit theorem. We can therefore use Eq. (3.1.4) to compute an equivalent BER from measured EVM as derived in [81][82],

$$\mathrm{BER} \approx \frac{1-L^{-1}}{\log_2 L} \operatorname{erfc}\left[\sqrt{\frac{3\log_2 L}{\left(L^2-1\right)\left(k\,\mathrm{EVM_m}\right)^2 \log_2 M}}\right]. \qquad (3.1.4)$$

Here L denotes the number of signal levels in each dimension (e.g. $L_{16\mathrm{QAM}} = 4$), and $\log_2 M$ is the number of bits encoded into each symbol. The conversion factor k is used to convert EVM normalized by the outermost constellation point to the EVM defined by the average power. For 256QAM we find $k^2 = 45 / 17$, and for 512QAM $k^2 = 529 / 165$. Unfortunately we spotted a misprint in [81], Eq. (4) where $\sqrt{2}$ has to be replaced by a 1. However, the corresponding plot in [81], Fig. 3(b) has been calculated correctly.

Conclusion: In this paper we show for the first time the transmission of 54 Gbit/s within an optical bandwidth of only 3 GHz over 44 km of ULAF employing zero roll-off Nyquist pulse shaping. Furthermore, we demonstrate signals with a data rate of 48 Gbit/s and 3 GHz optical bandwidth transmitted over 150 km of SSMF while the achieved signal quality is still suitable for error-free transmission using state-of-the-art FEC. Negligible inter-channel crosstalk is expected when employing the demonstrated signals in Nyquist WDM networks as the out-of-band suppression is >36dB. This will lead to a potential SE of 18 bit/s/Hz (16 bit/s/Hz) without and a net SE of 14.4 bit/s/Hz (15 bit/s/Hz) with FEC overhead. Corrections with signals having $\text{BER} \leq 3\times10^{-3}$ require an FEC with 6.7 % overhead [156], but for $\text{BER} \leq 1.9\times10^{-2}$ FEC algorithms introducing 25 % overhead [132] are required. As a result we find that transmission of PDM-256QAM signals leads to a higher net SE than transmitting PDM-512QAM data. The theoretical gain in SE of two bit per symbol is consumed by a decreased noise tolerance and by an increased FEC overhead.

Further discussions: The relation between EVM and BER [81][82] is theoretically well understood and has been used long since in wireless communications. EVM and BER are mathematically correlated in systems dominated by additive Gaussian amplifier noise. In optical communication systems where signals are made up from a large number of symbols, the reach is typically limited by additive Gaussian noise from the amplifiers. Nonlinear noise is not likely to dominate – otherwise the small distance in field strength from one symbol to the other would no longer allow transmission over multiple amplifier stages. However, even in the presence of nonlinear effects, where neighboring optical channels cross-couple random signals to the channel under consideration, these nonlinear perturbations are effectively Gaussian in nature due to the central limit theorem as has been demonstrated in extended simulations of WDM systems with up to 32 channels [157], Fig. 7. This is supported by findings that for narrowly spaced differentially phase shift keyed (DPSK) WDM systems, cross-phase modulation (XPM) contributes dominantly. However, XPM also adds noise with Gaussian statistic [158]. These ascertainments are supported by a number of experimental studies [79][31][81][82], where we proved the viability of this metric in optical communications. Even if there are some non-linear impairments degrading the signal quality such as phase noise or non-linear DSP, effects by Gaussian noise are usually dominant when measuring BER. As a consequence, soft-decision FEC can be applied to improve a raw BER of up to 1.9×10^{-2} to an effective BER of 1.1×10^{-15}, at the expense of a 25 % overhead [132]. In addition, recording complex error vectors allows to check the distribution of the noise [116] giving deeper insight into the nature of possible deficiencies, whereas BER measurements do not reveal any statistical information in that context.

5.2 Nyquist Frequency Divison Multiplexing

As an alternative to OFDM, where data is transmitted on numerous subcarriers (SC) with overlapping sinc-shaped spectra, Nyquist signals with numerous rectangular spectra can be multiplexed electrically. In the following section it is shown how these Nyquist frequency division multiplexed (NFDM) signals are transmitted optically. Furthermore the implications of limited processing power and real-world HW on the NFDM signal performance is discussed.

Nyquist Frequency Division Multiplexing for Optical Communications

R. Schmogrow, S. Wolf, B. Bäuerle, D. Hillerkuss, B. Nebendahl, C. Koos, W. Freude, and J. Leuthold

in Proc. Conf. on Lasers and Electro-Optics, paper CTh1H.2 (2012). [C20]

Asynchronously multiplexing various tributaries on a flexible grid and with the smallest possible guard bands is of highest interest in optical communications. While in the past most efforts were confined to reduce the guard bands in wavelength division multiplexed (WDM) schemes, spectral efficiency remained an issue. New digital signal processing (DSP) helped, though, to achieve record spectral efficiencies [118] by pulse shaping or filtering. Additionally, orthogonal frequency division multiplexing (OFDM) has emerged as a promising candidate in both, long-haul and access networks, offering high spectral efficiency and dynamic bandwidth allocation [23]. Compensation of accumulated chromatic dispersion (CD) can easily be dealt with by using a cyclic extension of the symbols (cyclic prefix, CP). However, drawbacks such as a CP-induced reduction of the symbol rate and the need for precise subcarrier spacing limit the application area. Recently, Nyquist WDM (NWDM) was introduced [24] using modulated sinc-shaped (Nyquist) pulses leading to rectangular spectra.

Here, we propose electrical Nyquist frequency division multiplexing (NFDM) where a broad spectrum is subdivided by modulating a large number of (possibly non-equidistant) electrical subcarriers with Nyquist pulses. Each of the resulting non-overlapping spectra could have a different width and represents one electrical NFDM channel. This is in contrast to OFDM, where a channel consists of a group of sinc-shaped overlapping equidistant spectra with identical width. Gaps between NFDM channels can accommodate pilot tones for carrier phase recovery and equalization. Like OFDM, NFDM supports variable channel loading (i.e. choosing the channel modulation format individually) and has a high dispersion tolerance inside each subcarrier band. As an advantage, no CP is needed for CD compensation. In this paper we demonstrate electrically generated NFDM consisting of up to 5 channels. To illustrate the versatility of the concept, we use both, a constant and a flexible subcarrier grid. All Nyquist channels carry 16QAM data. We study the influence of limited filter lengths and the impact of digital-to-analog converter (DAC) resolution.

5.2.1 Experimental Setup

The NFDM waveforms are generated by an Agilent M9180A arbitrary waveform generator (AWG) with a sampling rate of 12 GSa/s and a resolution of 12 bit, based on pre-computed and stored data (over-sampling factor 1.6). The AWG drives an optical I/Q-modulator which modulates an external cavity laser (ECL), Fig. 5.14(a). An erbium doped fiber amplifier (EDFA) boosts the signal which is eventually received by an Agilent N4391A optical modulation analyzer (OMA). Supported by external Matlab processing, the OMA decodes the data and evaluates their quality by the error vector magnitude (EVM) [81][82]. Sinc-shaped Nyquist pulses on three different electrical carrier frequencies and their associated rectangular power spectra are displayed in Fig. 5.14(b).

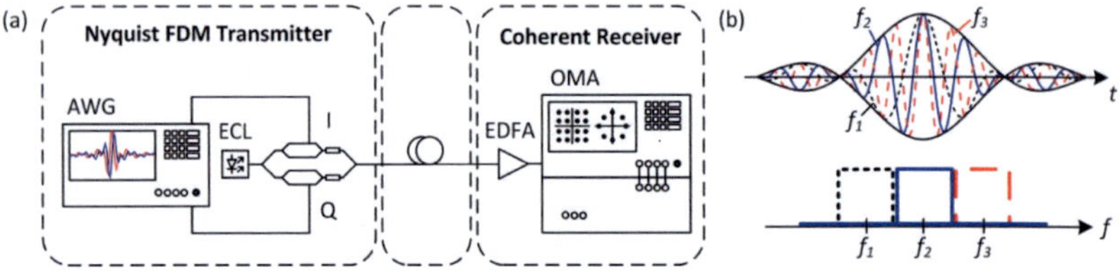

Fig. 5.14 Optical transmission of electrically generated NFDM signals. (**a**) Transmitter with Agilent M9180A AWG, ECL, and optical I/Q-modulator. Coherent receiver with Agilent N4391A OMA and EDFA pre-amplifier evaluates the EVM. The OMA is adapted to NFDM reception by external Matlab processing. (**b**) Nyquist sinc-pulses at three different electrical carrier frequencies and associated power spectra.

5.2.2 Results

Several 16QAM modulated NFDM signals (symbol length $T_s = 0.67$ ns corresponding to a symbol rate of 1.5 GBd) varying in sinc-pulse window width and resolution were generated and stored in the AWG. Multi-channel and single-channel spectra measured for optical back-to-back transmission are depicted in Fig. 5.15(a)–(c). Finite impulse response (FIR) filters of length $l = 512 \times T_s$ define the sinc-pulse window width. The NFDM spectrum in Fig. 5.15(a) represents five simultaneous Nyquist channels. The individual channel spectra are scaled to the maximum range of the AWG, see Fig. 5.15(b). All channels occupy the same bandwidth in an equidistant grid. Unlike OFDM signals where all subcarriers are arranged on an equidistant frequency grid, NFDM also supports flexible grids as the channel bandwidth may vary, Fig. 5.15(c). Symbol time windows and carriers in different NFDM channels need not be synchronized as opposed to OFDM. No significant crosstalk is observed even without the presence of a frequency guard band.

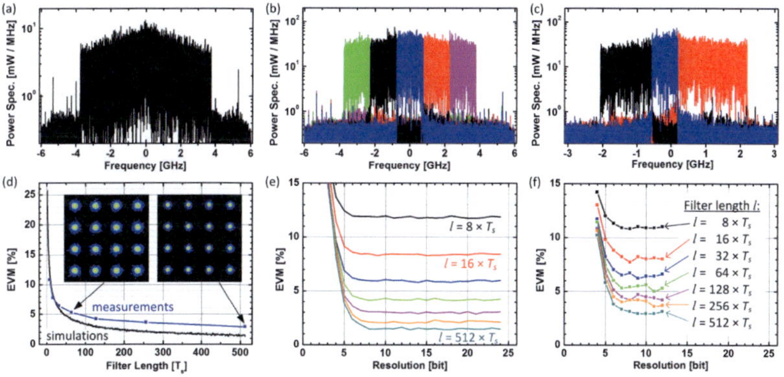

Fig. 5.15 Measured electrical NFDM power spectra generated with finite impulse response (FIR) filters. For each individual measurement the signal is scaled to the maximum range of the AWG. In Fig. 2(a)–(c), the filter length in terms of the symbol duration T_s is fixed to $l = 512 \times T_s$. (**a**) Composite power spectrum for five NFDM channels showing roll-off by finite sampling window and AWG bandwidth. (**b**) Five individually measured equidistant NFDM channels plotted in one graph. (**c**) Three individually measured channels located on a flexible grid and plotted in one graph. (**d**) Measured and simulated linear crosstalk between NFDM channels depending on filter length l. Crosstalk negligible if $l > 64 \times T_s$. (**e**) Simulated EVM due to DAC resolution only for various filter lengths l. (**f**) Measured EVM for various DAC resolutions and filter lengths l.

Linear inter-channel crosstalk is observed if neighboring spectra overlap due to insufficient length of the FIR filters [79]. In Fig. 5.15(d) the EVM of simulated and measured signals corresponding to different filter lengths are to be seen. The EVM is depicted for the channel with carrier frequency zero (Fig. 5.15(b)) as its quality is slightly inferior to the outer channels which have only one neighbor each. Crosstalk is negligible for filter lengths $l > 64 \times T_s$.

The influence of a limited DAC resolution on the signal quality is depicted in Fig. 5.15(e) (simulation) and Fig. 5.15(f) (measurement). We find that even for very long filters ($l = 512 \times T_s$) and very good channel isolation the effective resolution for 16QAM signals can be as low as $r = 7$ bit. Thus the proposed NFDM scheme performs well even with high-speed DACs that usually offer a resolution of 6 bit $\leq r \leq$ 8 bit. Measured EVM, Fig. 5.15(f), and simulated EVM, Fig. 5.15(e), do agree well.

5.3 OFDM and Nyquist Pulse-Shaping for Access Networks

There are several fields of application for optical transmission links. Among them one finds everything from ultra long-haul links, core and metro networks up to access networks and inhouse and data center communications. Depending on the scenario, different criteria or parameters speak for or against a certain realization. Since both pulse-shaping techniques, namely OFDM and Nyquist signals, are promising ways to achieve spectrally efficient networks, they will be investigated and discussed in the upcoming section for the use in an optical access network.

5.3.1 Uplink

Uplink Solutions for Future Access Networks

R. Schmogrow, P.C. Schindler, D. Hillerkuss, C. Koos, W. Freude, and J. Leuthold

in Access Networks and In-house Comm., paper AW4A.1 (2012). [C21]

Commercial optical access networks rely on on-off keyed (OOK) optical time division multiplexing (OTDM) and operate up to 10 Gbit/s per wavelength channel. Recently, orthogonal frequency division multiplexing (OFDM) [23] was investigated for access networks [138]. Higher data rates than conventional systems are expected together with an improved granularity [161]. As an alternative to OFDM, transmission of sinc-shaped Nyquist pulses has emerged as a competing technology with comparable or even better spectral efficiency [79]. OFDM signals comprise several frequency-locked subcarriers (SC) whose spectra are sinc-shaped and equidistant. Nyquist sinc-pulse modulation provides signals with rectangularly shaped spectra which can be spectrally multiplexed either in the electronic domain [152] or in the optical domain [121]. In contrast to OFDM, spectrally multiplexed Nyquist pulses neither have to be frequency-locked nor equidistant in their carrier frequencies.

Conventional OTDM optical network units (ONU) must process the wide-bandwidth signals in full, so high digital signal processing (DSP) power is required. In contrast, OFDM and multicarrier sinc-shaped Nyquist pulses [152] can be separated into groups of SCs, only one of each needs to be processed at the ONU.

In this paper we compare the performance of OFDM and sinc-shaped Nyquist pulse transmission both in an optical bandwidth of 5 GHz. Taking into regard a cyclic prefix and pilot tones, OFDM quadrature phase shift keyed (QPSK) signaling results in a net data rate of 7.5 Gbit/s, while sinc-shaped Nyquist QPSK transmission allows a net data rate of 10 Gbit/s.

5.3.1.1 Network Scenario

Individual stabilized laser sources in each ONU are costly but required for OFDM transmission, therefore we propose to seed all ONUs in a network segment by a laser source located at the optical line termination (OLT), see Fig. 5.16. The seed is sent along the feeder fiber, and then split by N to be amplified and modulated within each ONU. An electrical complex modulator converts either a single (or a group of) sinc-shaped OFDM SC, or, alternatively, a spectrally rect-shaped Nyquist channel to an electrical intermediate frequency (e.g. f_1). Electro-absorption modulators (EAM) are employed due to their polarization insensitivity and their high bandwidth. Finally, an optical filter suppresses the seed laser line to avoid back-scattering before the amplified data is sent back to the OLT. The signals of all ONUs superimpose at the splitter and form the aggregate uplink signal. This signal either comprises synchronized OFDM SCs, or (potentially asynchronous) Nyquist channels. The ONU receives the signal coherently.

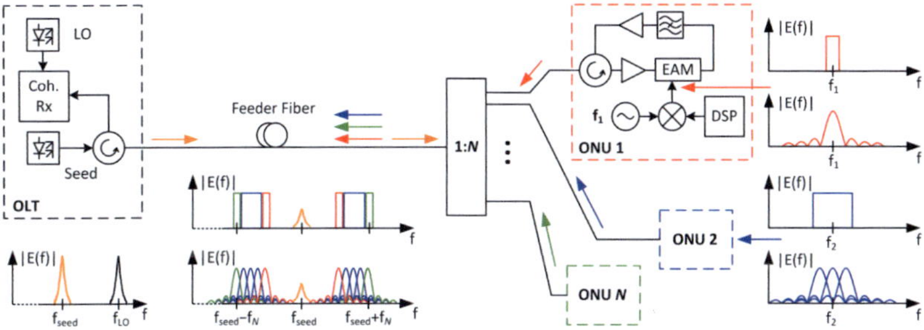

Fig. 5.16 Seeded uplink scenario for OFDM or sinc-shaped Nyquist pulse transmission. The OLT sends a seed laser line down the feeder fiber and provides the optical LO for a number of N ONUs. Each ONU modulates the amplified seed with an EAM. Electrical complex mixers drive the EAM. The mixers are fed with either a single (or a group of) OFDM SC, or, alternatively with rect-shaped Nyquist spectra. An optical bandstop removes the LO line to avoid back-scattering. All amplified uplink signals superimpose at the splitter and are coherently received by the OLT.

5.3.1.2 Experimental Setup and Results

The experimental setup of the seeded uplink along with characteristic optical spectra and received constellation diagrams is depicted in Fig. 5.17. The experiment comprises one OLT and two ONUs. In the OLT, an external cavity laser (ECL) is amplified by a semiconductor optical amplifier (SOA) and fed to a Mach Zehnder modulator (MZM). By driving the null-biased modulator with an 18.75 GHz tone, multiple lines are generated, two of which are separated by a Finisar waveshaper. One of the two carriers is polarization aligned and serves as the optical LO of the Agilent optical modulation analyzer (OMA). The other (seed) carrier is launched into the feeder fiber (standard single mode fiber, SSMF).

The seed is amplified by an SOA, which in this setup is shared among the two ONUs. In the network, however, each ONU provides an individual SOA for seed amplification. Next, the SOA output is split to serve both ONUs. An arbitrary waveform generator (AWG) generates real and imaginary parts of the OFDM or, alternatively, the sinc-shaped Nyquist pulse signals which drive the inputs of electrical complex modulators. Each ONU occupies a 5 GHz frequency band with center frequencies located at 28.5 GHz and 35 GHz. The data rate of the OFDM signal amounts to 7.5 Gbit/s per ONU, while the sinc-shaped Nyquist pulse signals carry 10 Gbit/s of data per ONU. The electrically up-converted signals are amplified and drive an EAM each. A 10 km fiber span is used to de-correlate data of both ONUs. The optical signal is then amplified by another SOA in each ONU. After combining the optical signals of both ONUs, a fiber Bragg grating (FBG) removes the seed to avoid signal degradation caused, e.g. by Rayleigh back-scattering. An optical isolator suppresses any counter-clockwise oscillation which could build up in the loop formed by the shared ONU and the parallel arrangement of ONU I and II. The aggregate uplink signal is sent to the OLT using the same feeder fiber, which also carries potential downstream data. An erbium doped fiber amplifier (EDFA) boosts the signal before reception by the OMA. Received constellation diagrams for back-to-back (b2b) and for 75 km of feeder fiber are shown as an inset in Fig. 5.17. Sinc-shaped Nyquist pulse constellations are colored blue, and OFDM-related constellations are colored red. We find similar performance for both types of modulations. However, some of the OFDM subcarriers are more degraded than the one depicted in Fig. 5.17 which in this scenario indicates an advantage of the sinc-shaped Nyquist pulse modulation over OFDM.

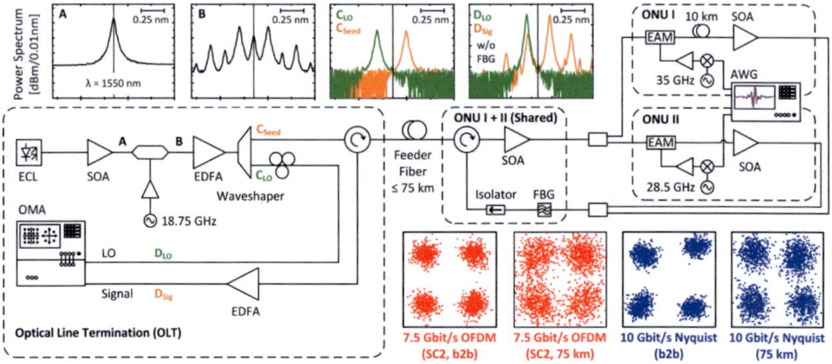

Fig. 5.17 Experimental setup and results. The seed carrier for the ONU and the LO for the OLT is generated using a continuous wave laser source (**A**) and a null-biased MZM (**B**). Both lines are separated by a waveshaper (**C**). The seed is then sent through up to 75 km of SSMF. Two ONUs share a first SOA which amplifies the seed. A splitter supplies the seed to the EAM in each of two ONUs where the OFDM or sinc-shaped Nyquist pulses are modulated onto the optical carrier. A 10 km spool of SSMF is used to de-correlate the signals of the two ONUs. A second SOA amplifies the optical signals before they are combined. The seed is suppressed by a fiber Bragg grating (FBG). An isolator prevents an amplification loop formed by the shared ONU and the parallel arrangement of ONU I and II. The aggregate uplink signal is then sent back to the OLT via the feeder fiber. At the OLT, the signal is amplified and received coherently (**D**) with the OMA. The inset shows constellation diagrams for QPSK obtained from OFDM (red) and sinc-shaped Nyquist pulse signals (blue) after back-to-back (b2b) and 75 km feeder fiber transmission.

5.3.2 Downlink

Remote Heterodyne Reception of OFDM-QPSK as Downlink-Solution for Future Access Networks

P. Schindler, R. Schmogrow, D. Hillerkuss, M. Nazarathy, S. Ben-Ezra, C. Koos, W. Freude, and J. Leuthold

in Access Networks and In-house Comm., paper AW4A.3 (2012). [C24]

Next-generation access networks rely on passive optical networks (PONs). State-of-the-art PON technology is presently limited to 10 Gbit/s on-off keying (OOK). However, to increase the spectral efficiency in future networks, complex modulation schemes like *M*-ary quadrature amplitude modulation (QAM) and dense wavelength division multiplexing in form of optical frequency division multiplexing (OFDM) [23] are required. These modulation formats and multiplexing schemes can be handled thanks to software defined multi-format transmitters [73][39]. The use of OFDM offers the benefit of having hundreds of densely packed subcarriers (SC) serving the same amount of subscribers per channel without having to use additional multiplexing formats like time division multiplexing (TDM) [138]. In order to increase the overall bandwidth of an OFDM-based PON one can utilize adaptive temporal allocation of frequencies when certain subscribers are not using their assigned capacity [161].

Future subscriber units (optical network unit, ONU) need to be designed colorless to meet budget-restrictions so that the same hardware can be used for all wavelengths. For best reception of complex modulation formats a coherent detection scheme is mandatory, so a local oscillator is needed. Rather than having an expensive laser at each and every ONU, this local oscillator is often distributed by the central office. Promising schemes employ reflective ONUs with a semiconductor optical amplifier (SOA) to provide optical pre-amplification for detection and boosting of the upstream (US) signal [162]. In this paper, we demonstrate a remote heterodyne approach, which exploits the mixing of the signal with an auxiliary carrier sent by the optical line termination (OLT).

5.3.2.1 Network Scenario

In typical optical access network Fig. 5.18(a), a single OLT provides a broadband downstream (DS) signal that is broadcast and received by each ONU. The DS signal is a 25 GHz OFDM signal, composed of 256 SCs including 16 pilot-tones, Fig. 5.18(b), offering each subscriber one or more SCs. The roll-off seen in the signal spectrum is primarily due to the bandwidth-limitations of the transmitter [39]. While every ONU optically receives the whole spectrum with a broad-band photodiode, it electrically demodulates only a portion. This is realized by electrical mixers converting only the desired SCs to the baseband, Fig. 5.18(c), thus enabling low-speed digital signal processing (DSP). The DS signal comprises 256 QPSK-modulated OFDM SCs, has no cyclic prefix, and is augmented with 16 pilot tones. As pilot tones, DC SC and the Nyquist tone do not carry data, a gross data rate of 46.5 Gbit/s is transmitted within the 25 GHz bandwidth. Due to the mentioned bandwidth constraints each ONU receives only 11.625 Gbit/s.

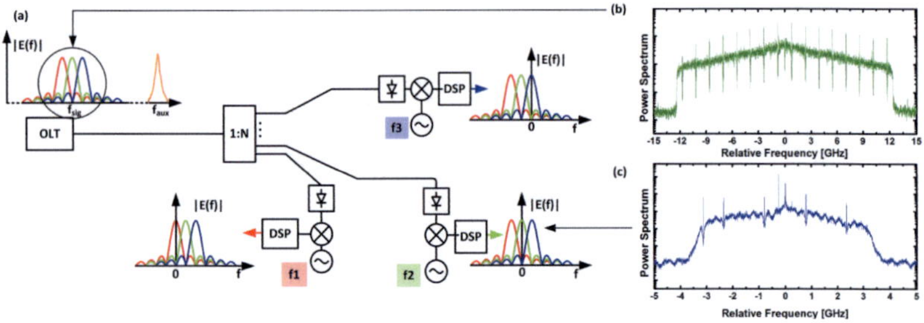

Fig. 5.18 Operation principle of the envisioned future access network **(a)** Network scenario. The OLT generates an OFDM-signal with 25 GHz bandwidth, which is composed of 256 subcarriers including 16 pilot-tones. Each subscriber in the PON has one or more SC assigned that can be selected and down-converted to the baseband by electrical mixers, thus receiving only this part of the information which is targeted for the individual subscriber. The mixer limits the overall receiver bandwidth to 6 GHz, so that the received SC number is reduced to a maximum of 64. **(b)** Measured spectra of the broadband OFDM signal as generated by the OLT. **(c)** Maximally wide received spectrum with 64 SC.

5.3.2.2 Experimental Setup and Results

For our investigations we implement a PON with one OLT and two ONUs to demonstrate the remote heterodyne reception scheme, see Fig. 5.19: The boosted output of an external cavity laser (ECL, spectrum A) is modulated with a Mach-Zehnder modulator, which is biased at the null point and driven with an 18.75 GHz sinusoidal. The output is filtered and split in a waveshaper to isolate two carriers, spaced by 37.5 GHz (spectrum B). One line serves as a carrier for DS modulation, while the second auxiliary carrier is used as a remotely generated local oscillator for heterodyne reception. The modulated signal and the auxiliary carrier (spectrum C) are boosted and launched into a fiber with a length of up to 100 km. After being split to feed the various ONUs, the optical signal is preamplified by an SOA prior to reception. A 2 nm filter suppresses amplified spontaneous emissions of the SOA. In a colorless ONU scheme this filter must be either wide enough to transmit all operating wavelengths (e.g. 10 nm), or it must be tunable. As a check if additional electrical amplification is needed, ONU I includes an extra electrical preamplifier. The optical power into both ONUs depends only on the different fiber lengths. The experimentally obtained constellation diagrams for one exemplary SC per transmission path length are shown in the lower right inset of Fig. 5.19. In the back-to-back case no extra electrical amplification is needed. However, for bridging up to 100 km of SMF-28, an additional electrical amplifier, as depicted in ONU I, is mandatory. Since no additional digital signal processing was applied (compensating e.g. for mixer imbalances) further improvement is expected.

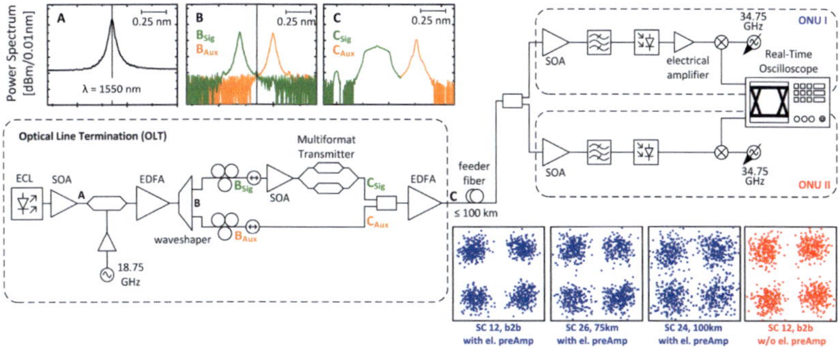

Fig. 5.19 Network scenario. The OLT comprises an external cavity laser (ECL) with optical booster (SOA) followed by a Mach-Zehnder modulator that generates two carriers with 37.5 GHz spacing. The two carriers are amplified and split by a waveshaper, the two output spectra of which are superimposed in Spectrum B. One carrier is modulated in the multi-format transmitter [73][39] to create the optical OFDM-signal Fig. 5.18(b). The second carrier is used as a remotely supplied local oscillator for heterodyne reception. Downstream and auxiliary carrier are combined by a 3dB coupler and launched into different numbers of SMF-28 spans. For this demonstration, two ONUs are realized with a heterodyne reception scheme: The signal is amplified in an SOA prior to reception. The superposition of signal and remotely supplied LO is detected by a single photodiode and electrically down-converted to the baseband with an electrical local oscillator for selecting the OFDM-subcarriers assigned to the respective ONU. In ONU I the electrical signal is amplified prior to demodulation, thereby improving the overall signal quality significantly.

5.4 OFDM with Raised Cosine Pulse Shape

Raised-Cosine OFDM for Enhanced Out-of-Band Suppression at Low Subcarrier Counts

R. Schmogrow, B. Bäuerle, D. Hillerkuss, B. Nebendahl, C. Koos, W. Freude, and J. Leuthold

in Signal Processing in Photonic Comm., paper SpTu2A.2 (2012). [C16]

Orthogonal frequency division multiplexing (OFDM) has emerged as a technology offering highly confined spectra and therefore high spectral efficiencies (SE) [23]. However, the number of subcarriers (SC) strongly influences the effective roll-off of the spectrum and therefore the suppression of out-of-band signal components. A large number of SCs leads to a roll-off that is insignificant compared to the total OFDM signal bandwidth [159], a fact which is useful when multiplexing several independent OFDM bands at a minimum spacing. A large number of SCs, however, calls for high processing efforts. It further reduces the signal's tolerance with respect to carrier phase noise [159] as the

OFDM symbol rate decreases with increasing SC counts, and the carrier coherence time must be maintained over a longer symbol period. On the other hand, Nyquist pulse shaping, a competing (but single carrier) technology, also produces signals with a potentially large SE and strong out-of-band suppression [79], which is infinite in the ideal case that sinc-pulses are transmitted. However, these pulses are free of intersymbol interference (ISI) only at specific sampling points und hence are sensitive to timing jitter. To lower this sensitivity and the computational complexity as well, sinc-like pulses with a raised-cosine (RC) spectrum [74] can be employed. The temporal impulse shape is determined by a roll-off factor $\beta > 0$ [74], which leads to a faster amplitude decay. These RC pulses still satisfy the Nyquist ISI criterion, i.e., they can be received without crosstalk from neighboring pulses.

In this paper we use RC-shaped OFDM symbols with sinc-like SC spectra, which are centered at the various SC frequencies,

$$\breve{h}_{\mathrm{N}}(f) = \frac{2\sin(\pi t / T_t)\cos(\pi\beta t / T_t)}{\omega_t t - (\omega_t t)^3 \beta^2 / \pi^2} \tag{3.1.5}$$

These RC-OFDM signals still require an infinite bandwidth according to Eq. (3.1.5) [74], but show a significantly increased out-of-band suppression even for a low number of SCs. Furthermore, in analogy to RC Nyquist pulses that become more tolerant towards time jitter with increasing β, the RC-OFDM signals become more robust with respect to frequency errors [160].

5.4.1 Experimental Setup

RC-OFDM waveforms are generated by an Agilent M8190A arbitrary waveform generator (AWG – 8 GSa/s, 14 bit), based on pre-computed data. The AWG drives an optical I/Q-modulator which modulates an external cavity laser (ECL) source, Fig. 5.20(a). An erbium doped fiber amplifier (EDFA) boosts the signal which is received by an Agilent N4391A optical modulation analyzer (OMA). Possible chromatic dispersion accumulated during transmission has to be compensated either inline or electronically. After reception, equalizing by inverse windowing is done with Matlab processing. The resulting data are fed back to the OMA which then decodes the symbols, and evaluates the signal quality by computing the error vector magnitude (EVM) [81][82]. Fig. 5.20(b) displays rect-shaped ($\beta = 0$, upper row) and RC-shaped ($\beta = 1$, lower row) OFDM symbols with three SCs along with the SC spectra.

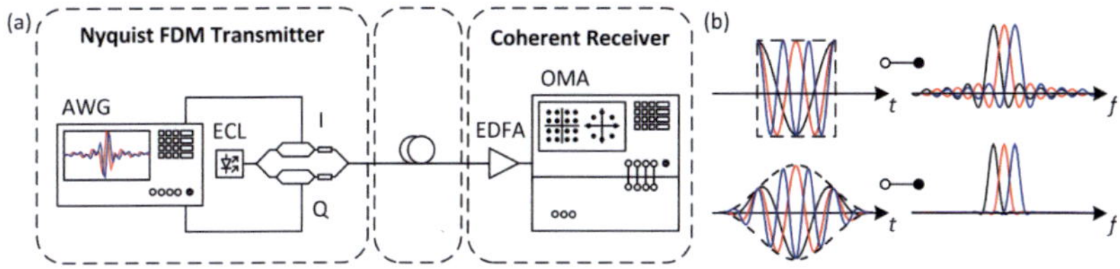

Fig. 5.20 Setup for RC-OFDM measurements with schematic of temporal pulse shapes and spectra. (**a**) An arbitrary waveform generator (AWG) holds pre-computed waveforms and drives an optical I/Q-modulator, which modulates an external cavity laser (ECL). The signals are amplified by an erbium doped fiber amplifier (EDFA) and received by the optical modulation analyzer (OMA). After reception, equalizing by inverse windowing is done with Matlab processing. The resulting data are fed back to the OMA which then decodes and displays the symbols. (**b**) Conventional OFDM ($\beta = 0$, upper row) and RC-OFDM ($\beta = 1$, lower row) are depicted schematically. Both signals comprise three subcarriers (black, red, and blue) which translate into corresponding true sinc-shaped spectra for OFDM, and three into sinc-like, more rapidly decaying spectra for RC-OFDM. The window functions are marked by dashed lines.

5.4.2 Results

We compare conventional and RC-OFDM comprising 60 SCs (generated with a 128 point IFFT) and two pilot tones for phase tracking. We vary the modulation of the SCs using quadrature phase shift keying (QPSK) and quadrature amplitude modulation (QAM) with 16 (16QAM) and 64 states (64QAM). The resulting symbol rate is 8 GBd / 128 = 62.5 MBd for $\beta = 0$. Measured and ensemble-averaged spectra for RC-OFDM with different roll-off factors β are depicted in Fig. 5.21(a)–(c) along with insets of typically received and windowed (dashed black) complex symbols (real/imaginary parts: red/blue). The noise floor is due to quantization noise generated by the receiver. The spectral decay enhances strongly when β increases, so that multiple frequency bands with RC-OFDM signals can be spaced more closely. The out-of-band power suppression at different distances from the main band for RC-OFDM signals with different β is shown in Fig. 5.21(d). The suppression inferred from simulated noiseless data (lines) is compared to actually measured suppression values (dots). For verifying the unchanged signal quality, we measure the error vector magnitude (EVM) for all formats and several β. When the roll-off β increases, the minimum required cyclic prefix (CP) increases as well. The relations between β, CP, and the symbol rate are seen in Fig. 5.21(f).

The out-of-band suppression is strongly enhanced when windowing conventional OFDM signals with a roll-off factor β. A steep spectral roll-off is achieved even for low SC counts (here: 60) and low $\beta = 0.3$. The signal quality is evaluated for QPSK, 16QAM, and 64QAM modulated SCs and proves to be virtually independent of the roll-off.

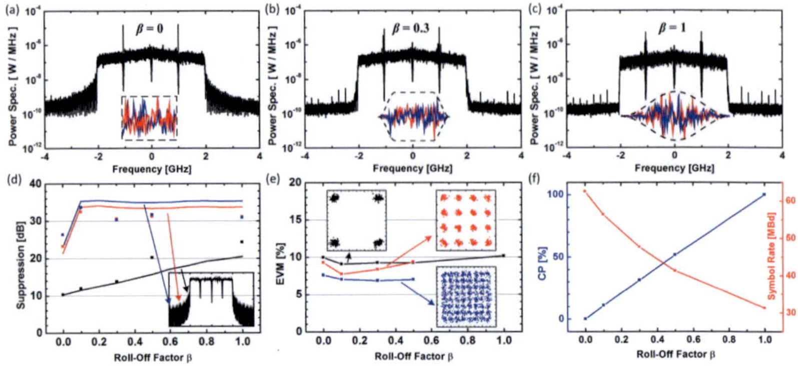

Fig. 5.21 Experiments with various roll-off factors β demonstrate superior out-of-band suppression for RC-OFDM. All signals comprise 60 SCs. Insets beneath the ensemble-averaged measured spectra (black) show typical received symbols with real (red) and imaginary (blue) parts. The window function is indicated by dashed lines. (**a**) Spectrum of conventional OFDM, (**b**) spectrum of RC-OFDM with $\beta = 0.3$, and (**c**) RC-OFDM spectrum with $\beta = 1$. (**d**) Simulated (lines) and measured (squares) out-of-band suppression for various roll-off factors and three different spectral distances from the main band (arrows). (**e**) Error vector magnitude (EVM) as a function of the roll-off factor and constellation diagram insets for QPSK, 16QAM, and 64QAM. (**f**) Minimum required cyclic prefix (CP) and associated symbol rates for different β.

Appendix A.

A.1. Quality Metrics (EVM, Q-Factor, SNR)

Quality metrics for advanced modulation formats in optical communications: OSNR, Q-factor, EVM, and BER

R. Schmogrow, B. Nebendahl, D. Hillerkuss, C. Koos, W. Freude, and J. Leuthold

in 13. ITG-Fachtagung Photonische Netze (2012). [C42]

Coherent optical transmission systems with advanced modulation formats such as *M*-ary quadrature amplitude modulation (*M*-QAM) are establishing quickly [1] as they offer superior performance compared to formats with direct detection, e.g., on-off keying (OOK). Instead of using format-specific modulators [2, 3], novel software-defined optical transmitters are more flexible. At the push of a button [4, 5], a large number of modulation formats can be chosen on the fly.

Especially for phase-sensitive transmission a quality metric different than for OOK signals is required. In laboratory experiments performed so far, most coherent receivers employ offline digital signal processing (DSP) at much reduced clock rates. While real-time oscilloscopes acquire data quickly, the data transfer to the processing unit is slow. This limitation leads to very time consuming bit error ratio (BER) estimations, especially if the signal quality is high: If a BER = 10^{-5} is to be determined with a reasonable statistical significance, at least 100×10^5 bits have to be recorded [13]. As a consequence, a faster — yet reliable — performance metric is needed, in particular for multi-channel systems [6, 7].

A.1.1. Error Vector Magnitude

Especially for phase-sensitive transmission a quality metric different than for OOK signals is required. In laboratory experiments performed so far, most coherent receivers employ offline digital signal processing (DSP) at much reduced clock rates. While real-time oscilloscopes acquire data quickly, the data transfer to the processing unit is slow. This limitation leads to very time consuming bit error ratio (BER) estimations, especially if the signal quality is high: If a BER = 10^{-5} is to be determined with a reasonable statistical significance, at least 100×10^5 bits have to be recorded [13]. As a consequence, a faster — yet reliable — performance metric is needed, in particular for multi-channel systems [6, 7].

$$\mathrm{EVM}_{\max} = \frac{\sigma_{\mathrm{err}}}{\left|E_{t,\max}\right|},\ \sigma_{\mathrm{err}}^2 = \frac{1}{N}\sum_{i=1}^{N}\left|E_{\mathrm{err},i}\right|^2, \qquad (6.1.1)$$
$$E_{\mathrm{err},i} = E_{r,i} - E_{t,i}\,.$$

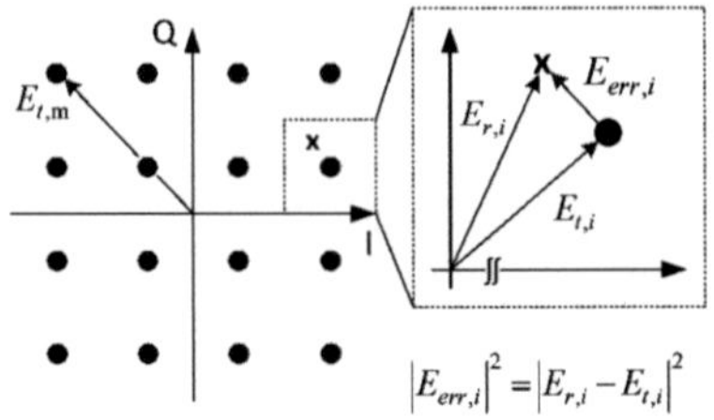

Fig. A 1 Constellation diagram and error vector for a 16QAM signal. Ideal constellation diagram with an actually transmitted value x. The blow-up illustrates the definition of the ith error vector $E_{\mathrm{err},i}$ in relation to the actually received signal vector $E_{r,i}$ and the vector $E_{t,i}$ of the transmitted signal. Modified from [9].

The power of the longest ideal constellation vector with magnitude $|E_{t,\max}|$ serves for normalization. Another definition found in the literature uses the average power $|E_{t,\mathrm{avg}}|^2$ of all M symbol vectors within a constellation leading to $\mathrm{EVM}_{\mathrm{avg}}$. The two EVM normalizations are related by a modulation format-dependent factor k,

$$\mathrm{EVM}_{\mathrm{avg}} = k\,\mathrm{EVM}_{\max},\ k^2 = \frac{\left|E_{t,\max}\right|^2}{\left|E_{t,\mathrm{avg}}\right|^2}, \qquad (6.1.2)$$
$$\left|E_{t,\mathrm{avg}}\right|^2 = \frac{1}{M}\sum_{i=1}^{M}\left|E_{t,i}\right|^2.$$

Table A.1 specifies the k-values relating the two discussed definitions for the modulation formats covered here.

Table A.1 Modulation format-dependent factor k^2 from Eq. (6.1.2)

	B/Q/8PSK	16QAM	32QAM	64QAM
k^2	1	9/5	17/10	7/3

Data aided and nondata-aided reception: In addition to the EVM normalization one has to discriminate between data-aided reception, where for measurement purposes the actually sent data are known, as opposed to nondata-aided reception, where the received data are unknown. The first case is standard for BER measurements where transmitted bits are directly compared with received bits, while the second case is more common for real-world receivers (disregarding, e.g., training sequences). For strongly noisy signals, nondata-aided reception tends to underestimate the EVM, because a received symbol could be nearer to a "wrong" constellation point than to its "right" position.

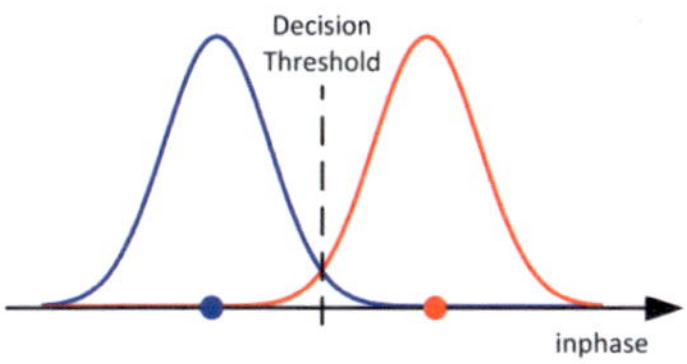

Fig. A 2 Schematic probability density functions (PDF) of error vectors. PDF of two neighboring symbols (blue and red) on the inphase axis for the real part of the electric field, along with their bell-shaped PDF of the error vectors. For nondata-aided reception, the required decision threshold could lead to false symbol assignments. Modified from [11].

In order to illustrate the influence of nondata-aided reception, Fig. A 2 shows the statistical properties of the received field vectors (belonging to two constellation points) with respect to their nominal positions. The deviations from each constellation point are the error vectors, which form a bell-shaped joint probability density function (PDF) as a function of real and imaginary parts of the error vector, centered at the respective constellation points. Fig. A 2 shows two neighboring constellation points for the inphase (real part) component of the electric field with the respective cross-section PDF in blue and red. We see that for nondata-aided reception, as is employed in this work, a decision threshold is required to assign received data to the closest constellation point. This can lead to false assignments and therefore to an underestimation of EVM and BER. In the following we investigate to what extend this fact has an influence on the BER estimation quality.

A.1.2. OSNR, EVM, BER and Q-Factor

Traditionally, the Q-factor metric is well established for optical systems with on-off keying (OOK). To estimate BER from Q, marks and spaces in the detected photocurrent are assumed to be superimposed with AWGN, the probability density of which is fully described by its mean and variance. A large Q leads to a small BER.

Unfortunately, the method cannot be simply transferred to QAM signals, where the optical carrier is modulated with multi-level signals both in amplitude and phase. Instead, the EVM is employed as described in the beginning of Section 2. If the received optical field is perturbed by AWGN only, i.e., if the PDF of Figure 2 are truly Gaussian, EVM can be related to BER and to the optical signal-to-noise ratio (OSNR) [9, 10]. A small EVM leads then to a small BER.

OSNR – When determining OSNR (measured for instance with an optical spectrum analyzer, OSA) the signal power is related to the noise in a fixed bandwidth 0.1 nm (corresponding to $B_{\text{ref}} = 12.5$ GHz at 1550 nm). However, for electrical SNR measurements the noise is measured in the same bandwidth B as the signal. With a signal power P_S and an unpolarized noise spectral density $w_{N,\text{ unpol}}$, the electrical signal-to-noise power ratio SNR can be related to the measured OSNR [12],

$$\mathrm{SNR} = 2\frac{P_S}{w_{N,\,\mathrm{unpol}}B}, \quad \mathrm{OSNR} = \frac{B}{2B_{ref}}\mathrm{SNR}. \tag{6.1.3}$$

EVM – The $\mathrm{EVM}_{\max}$ from Eq. (6.1.1) can then be estimated from SNR. The basic assumptions are AWGN, nondata-aided reception and quadratic M-QAM signal constellations. With Eq. (6.1.2) we find from [9, 10]:

$$\mathrm{EVM}_{\max} \approx \frac{1}{k}\left[\begin{array}{l}\dfrac{1}{\mathrm{SNR}} \\ -\sqrt{\dfrac{96}{\pi(M-1)\mathrm{SNR}}}\displaystyle\sum_{i=1}^{\sqrt{M}-1}\gamma_i\,\mathrm{e}^{-\frac{3\beta_i^2\mathrm{SNR}}{2(M-1)}} \\ +\dfrac{12}{M-1}\displaystyle\sum_{i=1}^{\sqrt{M}-1}\gamma_i\beta_i\,\mathrm{erfc}\left(\sqrt{\dfrac{3\beta_i^2\mathrm{SNR}}{2(M-1)}}\right)\end{array}\right]^{\frac{1}{2}}, \tag{6.1.4}$$

$$\gamma_i = 1 - i/\sqrt{M}, \quad \beta_i = 2i - 1.$$

The first term in Eq. (6.1.4), i.e., $\mathrm{EVM}_{\max} \approx 1/(k\sqrt{\mathrm{SNR}})$ or $\mathrm{EVM}_{\mathrm{avg}} \approx 1/\sqrt{\mathrm{SNR}}$, respectively, rewrites Eq. (6.1.1) and (6.1.2) for the case of data-aided reception and dominant AWGN. The remaining terms account for nondata-aided reception and disappear for large SNR. For large numbers of constellation points M only the first few terms in the summation need to be considered.

BER – To estimate a BER from $\mathrm{EVM}_{\max}$ we take into regard the number of bits $\log_2 M$ encoded into each QAM symbol of a quadratic constellation.

The BER for QAM signals can then be approximated by [8, 9] with the help of the complementary error function $\mathrm{erfc}(x) = 2/\sqrt{\pi}\int_x^{\infty}\mathrm{e}^{-t^2}\,\mathrm{d}t$,

$$\mathrm{BER} \approx \frac{2-2M^{-1/2}}{\log_2 M}\,\mathrm{erfc}\left[\sqrt{\frac{3}{2(M-1)(k\,\mathrm{EVM}_{\max})^2}}\right] \tag{6.1.5}$$

For Eq. (6.1.5), the same limitations as for Eq. (6.1.4) apply, but in this case data-aided reception is assumed.

For bipolar M-ary amplitude shift keyed (ASK) signals the BER can be approximately derived from EVM according to [12]

$$\mathrm{BER} \approx \frac{M-1}{M\log_2 M}\,\mathrm{erfc}\left[\sqrt{\frac{3}{(M^2-1)(k\,\mathrm{EVM}_{\max})^2}}\right] \tag{6.1.6}$$

Bipolar ASK with $M = 2$ is known as BPSK.

For unipolar ASK such as OOK the following relation holds [12]

$$\mathrm{BER} \approx \frac{1}{2}\,\mathrm{erfc}\left[\sqrt{\frac{1}{2\log_2 M(k\,\mathrm{EVM}_{\max})^2}}\right] \tag{6.1.7}$$

For phase shift keyed (PSK) signals, we find another expression [12] valid for $M > 4$ different phases,

$$\mathrm{BER} \approx \frac{1}{\log_2 M} \operatorname{erfc}\left[\sin\left(\frac{\pi}{M}\right) \Big/ \mathrm{EVM}_{\max} \right] \tag{6.1.8}$$

Q-factor – EVM and Q^2-factor are related. In direct detection OOK systems assuming electrical AWGN with a standard deviation $\sigma_1 \propto \sigma_0^2$ for the photocurrent $i_1 \propto |E_{t,\mathrm{m}}|^2$ of a mark, the Q-factor in the shot-noise limited case, $\sigma_0 \approx 0$, is defined in analogy to Eq. (6.1.1) by

$$\begin{aligned} Q &= \frac{i_1}{\sigma_1 + \sigma_0} \approx \frac{\left|E_{t,\max}\right|^2}{\sigma_Q^2} \approx \mathrm{SNR} \approx \frac{1}{\mathrm{EVM}_{\max}^2}, \\ \sigma_Q^2 &= \frac{1}{N} \sum_{i=1}^{N} \left[\left|E_{r,i}\right|^2 - \left|E_{t,i}\right|^2 \right] \end{aligned} \tag{6.1.9}$$

The Q^2-factor represents the electrical signal-to-noise power ratio SNR. For OOK signals it provides a good BER estimate [12], BER $\approx (1/2)\,\mathrm{erfc}(Q / \sqrt{2})$, see also Eq. (6.1.7) for $M = 2$. The EVM, however, is based on electrical fields and thus assesses the BER for a variety of formats, including phase sensitive formats. In the following, we compare the theoretical predictions Eq. (6.1.4)–(6.1.7) with numerical simulations and measurements.

A.1.3. Experimental setup and results

We measure OSNR, EVM and BER in a software-defined real-time multi format transmitter setup [4], Fig. A 3. We sequentially generate the six complex modulation formats B/Q/8PSK and 16/32/64QAM at symbol rates of 20 GBd and 25 GBd on an external cavity laser (ECL) at 1550 nm. The modulated carrier is kept at a fixed average power and combined with a variable noise source (amplified spontaneous emission (ASE) source with attenuator) to vary the OSNR. An optical spectrum analyzer (OSA) determines the amplified signal's OSNR. The symbol rate of 25 GBd occupies an optical bandwidth of 25 GHz, which happens to correspond to 2 B_{ref}. In this case, OSNR = SNR holds. An Agilent modulation analyzer (OMA) decodes the modulation and measures EVM and BER. The insets of Fig. A 3 display the spectrum of the ASE source as well as constellations at 25 GBd for three modulation formats and for the best OSNR available.

In Fig. A 4(a) we display the measured $\mathrm{EVM}_{\max}$ for various measured OSNR values (closed symbols for 20 GBd, open symbols for 25 GBd). The solid lines represent Eq. (6.1.4). For OSNR < 20 dB the theoretical prediction coincides with the measurement. Constellations of xQAM can be recovered for OSNR > 12 dB only. For OSNR > 20 dB the electronic noise dominates so that a change in OSNR does not influence the total SNR in Eq. (6.1.4). If the electronic noise contribution would be less, as is the case for systems with lower symbol rate and consequently smaller bandwidth, this error floor would be seen at higher OSNR only.

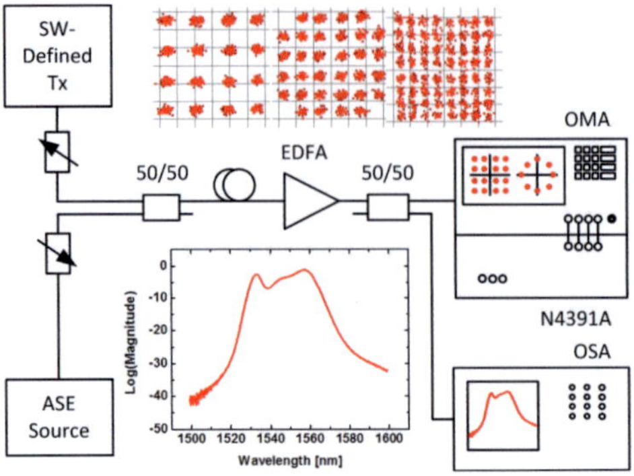

Fig. A 3 Experimental setup for BER and EVM measurements. The optical signal of a software-defined transmitter [4] generates a choice of six different modulation formats (three of which are shown) for an OSNR adjusted by injecting a variable amount of amplified spontaneous emission (ASE). After amplification with an erbium-doped fiber amplifier (EDFA), the OSNR is measured by an optical spectrum analyzer (OSA). The modulation is decoded by an Agilent optical modulation analyzer (OMA). Modified from [9].

Fig. A 4(b) shows the measured BER as a function of the measured EVM (closed symbols: 20 GBd, open symbols: 25 GBd). The solid lines represent Eq. (6.1.5) and (6.1.6), the dashed lines result from simulations.

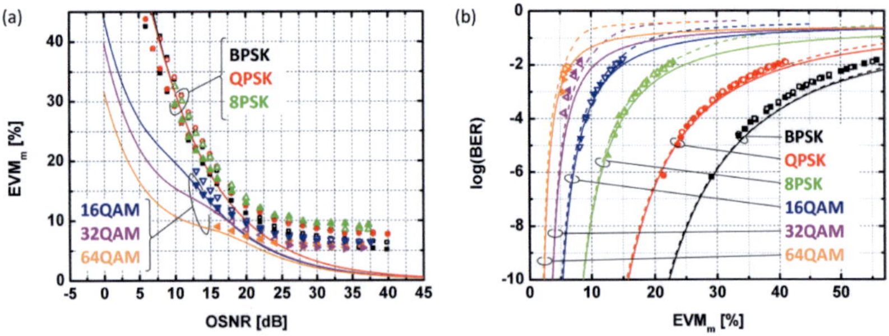

Fig. A 4 (**a**) Interdependencies of OSNR and EVM_m. Filled symbols represent measurements for a symbol rate of 20 GBd, open symbols for 25 GBd. Measured (symbols) and calculated [10] EVM_{max} (solid lines) as a function of OSNR. For high OSNR levels the measured plots have an error floor due to the electronic noise of the transmitter and receiver. The different error floors for Q/8PSK and xQAM stem from different factors k. The error floor for BPSK is lower because of transmitter specific properties. Modified from [9]. (**b**) Measured (symbols), simulated (dashed lines) and calculated [8 – 10] BER (solid lines) as a function of EVM_{max}. Modified from [9].

While measurement and simulation are based on nondata-aided reception, Eq. (6.1.5) and (6.1.6) assume data-aided detection. Still, measurement, analytical estimate and simulations coincide for a large range up to a BER of 10^{-2}.

Some more information can be extracted from Fig. A 4. While the 32QAM constellation is not strictly quadratic, it is nearly so, and hence the estimation quality is comparable to the one for the quadratic formats. The plots also show that the EVM depends on the format, as higher-order formats are more sensitive to noise than others, as predicted by Eq. (6.1.4)–(6.1.7).

A.2. Nyquist and OFDM Signaling: Useful Formulae

Real-time Nyquist pulse generation beyond 100 Gbit/s and its relation to OFDM

R. Schmogrow, M. Winter, M. Meyer, A. Ludwig, D. Hillerkuss, B. Nebendahl, S. Ben-Ezra, J. Meyer, M. Dreschmann, M. Huebner, J. Becker, C. Koos, W. Freude, and J. Leuthold

Optics Express 20(1), pp. 317–337 (2012). [J22]

At this point we want to describe in mathematical detail the properties of Nyquist signals and illustrate their close relation to OFDM. For a better understanding, Table A.2 presents an overview of frequently used symbols contrasting OFDM-specific to Nyquist-specific parameters. As usual, the symbols t and f stand for time and frequency.

Table A.2 Commonly used terms for OFDM and Nyquist signal description

OFDM		Nyquist	
T_s	temporal width of symbol	F_s	spectral width of symbol
F_s	spectral subcarrier spacing	T_s	temporal subcarrier spacing
f_k	spectral subcarrier position	t_k	temporal subcarrier position
$k = 0 \ldots N-1$	index for spectral position of subcarrier (sinusoidal in time)	$k = -\infty \ldots +\infty$	index for temporal position of subcarrier (sinusoidal in frequency)
$i = -\infty \ldots +\infty$	index for temporal position of symbol (rectangular in time)	$i = 0 \ldots N-1$	index for spectral position of symbol (rectangular in frequency)

For general usage we introduce a new set of variables z, Z, m, and Q since the equations can be related either to OFDM or Nyquist signals in frequency or time domain, whichever is of interest. First we define a rectangular window and a sinc-function by

$$\operatorname{rect}\left(\frac{z}{Z}\right)=\begin{cases}1 & \text{for}\,|z|<Z/2\\ 0 & \text{else}\end{cases},\quad \operatorname{sinc}\left(\frac{z}{Z}\right)=\begin{cases}1 & z=0\\ \dfrac{\sin(\pi z/Z)}{\pi z/Z} & z\neq 0\end{cases}. \tag{6.2.1}$$

In the following we summarize the mathematical relations that hold in general:

Orthogonality relations

$$\frac{1}{Z}\int_{-Z/2}^{+Z/2} \mathrm{e}^{\mathrm{j}2\pi mz/Z}\,\mathrm{e}^{-\mathrm{j}2\pi m'z/Z}\,\mathrm{d}z=\delta_{mm'} \qquad \text{for}\quad m,m'\in\mathbb{Z} \tag{6.2.2}$$

$$\frac{1}{Z}\int_{-\infty}^{+\infty} \operatorname{sinc}\left(\frac{z}{Z}-m\right)\operatorname{sinc}\left(\frac{z}{Z}-m'\right)\mathrm{d}z=\delta_{mm'} \qquad \text{for}\quad m,m'\in\mathbb{Z} \tag{6.2.3}$$

Series expansions

We expand functions $\varphi(z)$ in a series of orthogonal complex harmonics with the help of the orthogonality relation Eq. (6.2.2), and functions $\psi(z)$ in a series of orthogonal sinc-functions observing the orthogonality relation Eq. (6.2.3),

$$\varphi(z)=\sum_{m=-\infty}^{+\infty}\varphi_m\,\mathrm{e}^{\mathrm{j}2\pi mz/Z}, \tag{6.2.4}$$

$$\varphi_m=\frac{1}{Z}\int_{-Z/2}^{+Z/2}\varphi(z)\,\mathrm{e}^{-\mathrm{j}2\pi mz/Z}\,\mathrm{d}z \tag{6.2.5}$$

$$\psi(z)=\sum_{m=-\infty}^{+\infty}\psi_m\operatorname{sinc}\left(\frac{z}{Z}-m\right) \tag{6.2.6}$$

$$\psi_m=\frac{1}{Z}\int_{-\infty}^{+\infty}\psi(z)\operatorname{sinc}\left(\frac{z}{Z}-m\right)\mathrm{d}z \tag{6.2.7}$$

Power relations

$$\frac{1}{Z}\int_{-Z/2}^{+Z/2}|\varphi(z)|^2\,\mathrm{d}z=\sum_{m=-\infty}^{+\infty}|\varphi_m|^2, \tag{6.2.8}$$

$$\frac{1}{Z}\int_{-\infty}^{+\infty}|\psi(z)|^2\,\mathrm{d}z=\sum_{m=-\infty}^{+\infty}|\psi_m|^2. \tag{6.2.9}$$

Peak power of a sum of sinc-functions

Nyquist signals and OFDM spectra are both described by a sum $s(z)$ of equidistantly shifted sinc-functions, Eq. (6.2.6). We are interested in a worst-case estimation of the maximum power $|s_{\max}|^2$. To this end we assume a constant height of all sinc-functions by choosing coefficients $|\psi_m| = 1$ with equal magnitude. The signs of the coefficients ψ_m are then selected such that a maximum $s_{\max}(z_{\max})$ is found at some position $z_{\max}$. We start by expanding the special function $s^{(1)}(z)=1$ in a series of sinc-functions, Eq. (6.2.6). The expansion coefficients ψ_m are calculated to be $\psi_m = 1\,\forall m$ by evaluating Eq. (6.2.7) and observing that ([125], Vol. 1, p. 454, formula 3.721 1.)

$$\frac{1}{Z}\int_{-\infty}^{+\infty} \operatorname{sinc}\left(\frac{z}{Z}-m\right)\mathrm{d}z = 1. \tag{6.2.9}$$

From Eq. (6.2.6) it follows that

$$s^{(1)}(z) = \sum_{m=-\infty}^{+\infty} \operatorname{sinc}\left(\frac{z}{Z}-m\right) = 1. \tag{6.2.10}$$

Eq. (6.2.10) shows that performing a summation of equally spaced sinc-functions with identical weight leads to a value of 1 at any position z. This value can be exceeded by choosing the expansion coefficients ψ_m appropriately. For this it should be noted that the sinc-function flips sign between adjacent intervals bounded by zeros. The maximum value of the sum $s(z)$ is obtained when all sinc-functions have the same sign in the z-interval under consideration. This is true for

$$\begin{aligned} s(z) &= \sum_{m=-\infty}^{+\infty} \psi_m \operatorname{sinc}\left(\frac{z}{Z}-m\right) \\ &= \sum_{m=0}^{+\infty}(-1)^m \operatorname{sinc}\left(\frac{z}{Z}-m\right) + \sum_{m=-\infty}^{-1}(-1)^{m+1} \operatorname{sinc}\left(\frac{z}{Z}-m\right), \end{aligned} \tag{6.2.11}$$

where the coefficients ψ_m have been chosen such that pairs of sinc-functions $m = (0, +1)$; $(-1, +2)$; $(-2, +3)$; … have all a positive sign in the interval $0 < z < Z$. The resulting function $s(z)$ is monotonic in $0 < z < Z$ and symmetrical with respect to $z = Z/2$, so that the superposition of each pair has its maximum at this point, as will be explained in the following.

Consider a function $f(u)$ which is monotonic in an interval $-U < u < +U$ ($U > 0$). In this interval the sum $s_f(u) = f(u) + f(-u)$ has an extremum if $s_f'(u) = f'(u) - f'(-u) = 0$, i.e., for $u = 0$. This result as applied to Eq. (6.2.11) means that the maximum is found at the symmetry point $z_{\max} = Z/2$ of the sum $s(z)$,

$$s_{\max}\left(\tfrac{1}{2}Z\right) = \sum_{m=-\infty}^{+\infty} \left|\operatorname{sinc}\left(\tfrac{1}{2}-m\right)\right| \tag{6.2.12}$$

Note that the sum does not converge. However, Eq. (6.2.12) also applies to a finite sum with a maximum of Q sinc-functions, from which the maximum power $s^2_{Q,\max}(Z/2)$ can be computed,

$$s_{Q,\max}\left(\tfrac{1}{2}Z\right)=\sum_{m=-Q/2+1}^{Q/2}\left|\operatorname{sinc}\left(\tfrac{1}{2}-m\right)\right| \tag{6.2.13}$$

Average power of an oversampled sinc-function

For deriving the average power of a sum of oversampled shifted sinc-functions sinc($qz/Z-m$) (oversampling factor q), we expand $\psi(z)=\operatorname{sinc}(z/Z)$ Eq. (6.2.6), but this time in terms of oversampled sinc-functions sinc($qz/Z-m$). We find the expansion coefficients $\psi_m=\operatorname{sinc}(m/q)$ according to Eq. (6.2.7) and write

$$\psi(z)=\operatorname{sinc}\left(\frac{z}{Z}\right)=\sum_{m=-\infty}^{+\infty}\operatorname{sinc}\left(\frac{m}{q}\right)\operatorname{sinc}\left(q\frac{z}{Z}-m\right). \tag{6.2.14}$$

By substituting $\psi(z)$ in the power relation Eq. (6.2.9) and by applying the orthogonality relation Eq. (6.2.3) we find the average power

$$\bar{P}=\frac{1}{Z}\int_{-\infty}^{+\infty}\left|\psi(z)\right|^2\mathrm{d}z=\frac{1}{q}\sum_{m=-\infty}^{+\infty}\operatorname{sinc}^2\left(\frac{m}{q}\right)=1. \tag{6.2.15}$$

In real life, oversampling the base functions by a factor q (preferably $q=2$) is needed to simplify the filtering of a Nyquist channel. The sinc(z/Z)-function is then represented not by a number of Q base functions as in Eq. (6.2.13), but by $q\,Q$ base-functions, and again orthogonality is lost in the strict sense. Nevertheless we approximate Eq. (6.2.14) by

$$\psi(z)=\operatorname{sinc}\left(\frac{z}{Z}\right)\approx\sum_{m=-qQ/2+1}^{+qQ/2}\operatorname{sinc}\left(\frac{m}{q}\right)\operatorname{sinc}\left(q\frac{z}{Z}-m\right). \tag{6.2.16}$$

If $q\,Q$ is large enough, the average power should be still close to 1,

$$P^{(q,Q)}=\frac{1}{q}\sum_{m=-qQ/2+1}^{+qQ/2}\operatorname{sinc}^2\left(\frac{m}{q}\right)\approx 1. \tag{6.2.17}$$

In reality we not only have a finite number $q\,Q$ of base functions, but the so far assumed equal modulus for all expansion coefficients must be modified if QAM modulated signals come into play. In this case, the approximated average power Eq. (6.2.17) needs to be divided by a format dependent factor k^2 [81][82], which relates the maximum power of the constellation points to the mean power for all constellation points. Therefore we write approximately

$$\bar{P}\approx P^{(q,Q,k)}=\frac{P^{(q,Q)}}{k^2}=\frac{1}{k^2 q}\sum_{m=-qQ/2+1}^{qQ/2}\operatorname{sinc}^2\left(\frac{m}{q}\right)\approx 1. \tag{6.2.18}$$

PAPR for a Nyquist signal

The average power in Eq. (6.2.18) serves as reference for the PAPR whereas the maximum power is determined by Eq. (6.2.13). We obtain

$$\mathrm{PAPR}_{\mathrm{Nyquist}} = \frac{s^2_{Q,\max}\left(\frac{1}{2}Z\right)}{\overline{P}} = k^2 \frac{\left[\sum_{m=-Q/2+1}^{Q/2} \left|\mathrm{sinc}\left(\frac{1}{2}-m\right)\right|\right]^2}{\frac{1}{q}\sum_{m=-qQ/2+1}^{+qQ/2} \mathrm{sinc}^2\left(\frac{m}{q}\right)}. \tag{6.2.19}$$

This equation corresponds to Eq. (2.5.11) in the main body of this thesis.

Peak power of an OFDM symbol

An OFDM symbol with Q sinusoidal carriers constant within a window of width Z be given by

$$s(z) = \sum_{m=1}^{Q} \sqrt{2}\cos\left(2\pi m \frac{z}{Z} + \alpha_m\right). \tag{6.2.20}$$

If the phases α_m of the Q carriers are chosen accordingly and all symbols have maximum values, then all amplitudes add up leading to:

$$s_{Q,\max} = \sqrt{2}\,Q. \tag{6.2.21}$$

Average power of an OFDM symbol

The average power can be determined with the power relation Eq. (6.2.8),

$$\begin{aligned} P^{(Q)} &= \frac{1}{Z}\int_{-Z/2}^{+Z/2} \left|s(z)\right|^2 \mathrm{d}z = \sum_{m=1}^{Q} \frac{2}{Z}\int_{-Z/2}^{+Z/2} \cos^2\left(2\pi m \frac{z}{Z} + \alpha_m\right)\mathrm{d}z \\ &= \sum_{m=1}^{Q} \frac{2}{Z}\frac{1}{2} Z = \sum_{m=1}^{Q} 1 = Q. \end{aligned} \tag{6.2.22}$$

Strictly speaking, orthogonality is lost if the OFDM spectrum is truncated as is always the case in reality. Nevertheless, Eq. (6.2.22) represents a good approximation for the average power of an OFDM signal comprising a sufficient number of Q subcarriers. Similar to the arguments leading to Eq. (6.2.18), the average power in a symbol needs to be divided by a format dependent factor k^2 [81][82] such that the average power in a symbol is

$$\overline{P}^{(Q,k)} = Q / k^2 \tag{6.2.23}$$

PAPR of an OFDM symbol

The PAPR follows by relating Eq. (6.2.21) to Eq. (6.2.23). We find

$$\mathrm{PAPR}_{\mathrm{OFDM},k=1} = \frac{s^2_{Q,\max}}{\overline{P}^{(Q,k)}} = 2k^2Q \tag{6.2.24}$$

For Eq. (6.2.24) the same number of elementary functions was adopted as for Eq. (6.2.19)

Spectrum of a Nyquist signal

The spectrum $Y^{(0)}_{\mathrm{FIR}}(f,R)$ of a Nyquist signal having a finite extent in time results from convolving a rectangular spectrum $Y^{(0)}(f)$ of Eq. (2.5.3) (representing the spectrum symmetrical to $f = 0$ of an infinitely extended baseband Nyquist sinc-impulse) with a sinc-shaped spectrum $W(f,R)$ (representing the spectrum of a rectangular time window $w(t) = \mathrm{rect}[t\,/\,(RT_s\,/\,q)]$ which depends on the number of filter taps R and the over-sampling factor q),

$$Y^{(0)}_{\mathrm{FIR}}(f,R) = Y^{(0)}(f) * W(f,R) = T_s\,\mathrm{rect}\left(\frac{f}{F_s}\right) * \frac{R}{q} T_s\,\mathrm{sinc}\left(R\frac{f}{qF_s}\right). \tag{6.2.25}$$

On evaluation we find in terms of the sine integral [124] $\mathrm{Si}(z) = \int_0^z (\sin v\,/\,v)\,\mathrm{d}v$

$$Y^{(0)}_{\mathrm{FIR}}(f,R) = \frac{T_s}{\pi}\left[\mathrm{Si}\left(\pi R\frac{f + F_s/2}{qF_s}\right) - \mathrm{Si}\left(\pi R\frac{f - F_s/2}{qF_s}\right)\right]. \tag{6.2.26}$$

Glossary

Acronyms and Symbols

a	Real-part of a complex number in Eq. (2.4.44)
$A(t)$	Slowly varying amplitude of an optical carrier
AD	Analog-to-digital
ADC	Analog-to-digital converter
ASE	Amplified spontaneous emission
ASIC	Application specific integrated circuit
ASK	Amplitude-shift keying (format)
AWG	Arbitrary waveform generator
b	Imaginary-part of a complex number in Eq. (2.4.44)
BER	Bit-error ratio
BERT	Bit-error ratio tester
BRAM	Block random access memory
c	Speed of light
C	Chromatic dispersion parameter in Eq. (2.2.7)
C	Channel capacity in Eq. (3.1.2)
CD	Chromatic dispersion
c_{ik}	Complex modulation coefficient
CMA	Common modulus algorithm
CPA	Constant power algorithm
CP	Cyclic prefix
CPU	Central processing unit
CW	Continuous-wave
$d(n)$	Deviation (at discrete times n) from the ideal filter output in Eq. (2.4.42)
DA	Digital-to-analog
DAC	Digital-to-analog converter
DC	Direct current
DCA	Digital communications analyzer
DCF	Dispersion compensating fiber
DFB	Distributed feedback
DFT	Discrete Fourier transform
DGD	Differential group delay
DML	Directly modulated laser
DP	Dual polarization
DQPSK	Differential quadrature phase-shift keying (format)

DS	Downstream
DSP	Digital signal processing
$E[\cdot]$	Expectation value in Eq. (2.4.43)
E, $E(t)$	Electric field in Eq. (2.1.1)
e	Elementary charge (1.6×10^{-19}As) in Eq. (2.3.2)
$e(n)$	Error signal at discrete times *n* in Eq. (2.4.42)
EAM	Electro absorption modulator
ECL	External cavity laser
ECOC	European Conference on Optical Communications
EDFA	Erbium-doped fiber amplifier
EOM	Electro optic modulator
EQ	Equalizer
Eq.	Equation
EVM	Error vector magnitude
EVM_{avg}	EVM normalized to the average power of the constellation diagram
EVM_{max}	EVM normalized to the (max) power of the outermost constellation point
f	Frequency in Eq.
FEC	Forward error correction
FFT	Fast Fourier transform
FIFO	First in first out
FIR	Finite duration impulse response
f_k	Frequency of the *k*-th subcarrier (OFDM) in Eq. (2.5.1)
f_{max}	Maximum frequency component of a band-limited signal
FPGA	Field programmable gate array
f_s	Sampling frequency in Eq. (2.4.1)
F_s	Symbol rate (the inverse symbol duration $F_s = 1/T_s$) in Eq. (2.5.5)
$G(f)$	Channel frequency response in Eq. (2.2.8)
$g(t)$	Channel impulse response in Eq. (2.2.9)
GCD	Greatest common divisor
GVD	Group velocity dispersion
h_r	Filter coefficients
$\hbar$	Reduced Planck constant (1.055×10^{-34}Js) in Eq. (2.3.2)
HW	Hardware
I	Intensity
$\mathbf{I}$	Identity matrix
i, $i(t)$	Photo-current in Eq. (2.3.7)
ICI	Inter carrier interference
IDFT	Inverse discrete Fourier transform
IF	Intermediate frequency
IFFT	Inverse fast Fourier transform

IIR	Infinite duration impulse response
IM	Intensity modulation
$\Im\{\cdot\}$	Imaginary part in Eq. (2.4.14)
IO	Input output
IP	Internet protocol
IPI	Inter polarization interference
ISI	Inter symbol intereference
j	Imaginary unit in Eq. (2.1.1)
J	Cost function in Eq. (2.4.13)
$\vec{J}$	Jones vector in Eq. (2.2.11)
k	Propagation constant in Eq. (2.2.4)
k	Conversion factor between EVM normalizations (max–avg) in Eq. (3.1.4)
LC	Left-handed circular polarized light
$LiNbO_3$	Lithium Niobate
LMS	Least mean squares
LO	Local oscillator
LUT	Look-up table
M	Length of the modulation alphabet (e.g. *M*-ary QAM)
MGT	Multi gigabit transceiver
MIMO	Multiple input multiple output
MMA	Mean modulus algorithm
MMF	Multi-mode fiber
MPA	Mean power algorithm
MUX	Multiplexing, Multiplexer
MZI	Mach-Zehnder interferometer
MZM	Mach-Zehnder modulator
n	Refractive index in Eq. (2.2.1)
NF	Noise figure
NFDM	Nyquist frequency division multiplexing
NRZ	Non-return-to-zero
N-WDM	Nyquist wavelength division multiplexing
NYQ	Rectangular Nyquist filter
OFC	Optical fiber communications conference
OFDM	Orthogonal frequency division multiplexing
OLT	Optical line termination
OMA	Optical modulation analyzer
ONU	Optical network unit
OOK	On-off keying (format)
OSA	Optical spectrum analyzer
OSNR	Optical signal-to-noise ratio measured with 0.1 nm reference bandwidth

OTDM	Orthogonal time division multiplexing
P	Power in Eq. (2.5.9)
PAM	Pulse amplitude modulation
PAPR	Peak-to-average power ratio
PAR	Place and route
PBS	Polarization beam splitter
PC	Personal computer
PD	Photo detector
PDF	Probability density function
PDL	Polarization dependent loss
PDM	Polarization division multiplexing
PMD	Polarization mode dispersion
PMF	Polarization-maintaining fiber
POF	Plastic optical fiber
PON	Passive optical network
PPC	Power PC
PRBS	Pseudo random binary sequence
PSK	Phase shift keying (format)
PPLN	Periodically-poled Lithium Niobate
PRBS	Pseudo-random bit sequence
PSK	Phase-shift keying (format)
Q	Quadrature of a carrier
q	Oversampling factor in Eq. (2.5.5)
QAM	Quadrature amplitude modulation (format)
QPSK	Quadrature phase shift keying (format)
R	Responsitivity of a photo detector in Eq. (2.3.2)
R	Filter order in Eq. (2.5.5)
$\mathbf{R}$	Rotation matrix in Eq. (2.4.11)
RC	Raised cosine
RC	Right-handed circular polarized light
$\Re\{\cdot\}$	Real part in Eq. (2.4.14)
RRC	Root raised cosine (or square root raised cosine)
RTL	Register transfer level
Rx	Receiver
RZ	Return-to-zero
$s(z)$	Nyquist or OFDM signal in Eq. (2.4.22)
$\vec{s}$	Stokes vector in Eq. (2.2.12)
$\vec{S}$	Normalized Stokes vector Eq. (2.2.13)
SC	Subcarrier
SE	Spectral efficiency

SMF	Single mode fiber
SNR	Signal-to-noise ratio
SOA	Semiconductor optical amplifier
SOP	State of polarization
SSMF	Standard single mode fiber
SW	Software
t	Time in Eq. (2.1.1)
T	Time period in Eq. (2.4.4)
t_k	Temporal position of Nyquist pulse in Eq. (2.5.3)
T_s	Symbol period (the inverse symbol rate $T_s = 1/F_s$) in Eq. (2.5.10)
Tx	Transmitter
U	SOP transfer matrix of an optical fiber in Eq. (2.4.10)
ULAF	Ultra large area fiber
US	Upstream
v	Velocity
v, $v(t)$	Applied voltage
VHDL	Very high speed integrated circuit hardware description language
VOA	Variable optical attenuator
VSA	Vector signal analyzer
WDM	Wavelength division multiplexing
x	Signal on x-polarization
$X(f)$	OFDM spectrum in Eq. (2.5.2)
$x(t)$	OFDM signal in Eq. (2.5.1)
XGPIO	General purpose IO of Xilinx Virtex FPGAs
XPM	Cross phase modulation
y	Signal on y-polarization
$Y(f)$	Nyquist spectrum in Eq. (2.5.3)
$y(t)$	Nyquist signal in Eq. (2.5.4)
Z	Symbol rate F_s or period T_s in Eq. (2.4.23)
z	Time t or frequency f in Eq. (2.4.22)
z^{-1}	Unit delay
α	Rotation angle of the polarization in Eq. (2.4.11)
β	Roll-off factor of raised-cosine filters in Eq. (3.1.1)
$\beta(\omega)$	Propagation constant in Eq. (2.2.5)
δ	Dirac impulse in Eq. (2.4.2)
$\delta(\cdot)$	Correlation function in Eq. (2.4.51)
Δ	Difference (e.g. Δt) in Eq. (2.4.8)
η	Quantum efficiency in Eq. (2.3.2)
φ	Phase in Eq. (2.1.2)

$\mathbf{\Phi}$	Retardation matrix in Eq. (2.4.11)
$\Phi\{\cdot\}$	Angle of a complex number in Eq. (2.4.37)
λ	Wavelength in Eq. (2.2.3)
μ	Step size parameter in Eq. (2.4.52)
ν	Spectral offset in Eq. (2.4.23)
ψ	Angle to determine the state of polarization in Eq. (2.2.12)
σ	Standard deviation in Eq. (6.1.1)
τ	Temporal offset in Eq. (2.4.23)
ϑ_{uu}	Autocorrelation function in Eq. (2.4.49)
ϑ_{ud}	Cross-correlation function in Eq. (2.4.50)
θ	Phase in Eq. (2.4.5)
ω	Angular frequency ($2\pi f$) in Eq. (2.1.1)
χ	Angle to determine the state of polarization in Eq. (2.2.12)
ζ	Offset either in time ($\zeta = \tau$) or frequency ($\zeta = \nu$) in Eq. (2.4.23)

References

[1] "Cisco visual networking index: Forecast and methodology, 2011–2016," Cisco Systems, Inc., San Jose, CA, USA, white paper, May. 2012. [Online]. Available: http://www.cisco.com/en/US/solutions/collateral/ns341/ns525/ns537/ns705/ns827/white_paper_c11-481360.pdf

[2] L. J. Cimini, Jr. and G. J. Foschini, "Coding of optical on-off keying signals impaired by phase noise," IEEE Transactions on Communications 38(9), 1301–1307 (1990).

[3] N. J. Frigo, P. P. Iannone, P. D. Magill, T. E. Darcie, M. M. Downs, B. N. Desai, U. Koren, T. L. Koch, C. Dragone, H. M. Presby, and G. E. Bodeep, "A wavelength-division multiplexed passive optical network with cost-shared components," IEEE Photon. Technol. Lett. 6(11), 1365–1367 (1994).

[4] J. G. Smith "Spectrally efficient modulation", in Proc. IEEE Int. Conf. Commun. (ICC'77), 37–41 (1977).

[5] N. Kapany, "Fiber Optics. Part I. Optical Properties of Certain Dielectric Cylinders," J. Opt. Soc. Am. 47(5), 413–418 (1957).

[6] T. Okoshi, "Recent progress in heterodyne/coherent optical-fiber communications," J. Lightwave Technol. 2(4), 341–346 (1984).

[7] A.S. Karar, M. Yanez, Y. Jiang, J.C. Cartledge, J. Harley, and K. Roberts, "Electronic dispersion pre-compensation for 10.709-Gb/s using a look-up table and a directly modulated laser," Opt. Express 19, B81–B89 (2011).

[8] S. Ryu, S. Yamamoto, H. Taga, N. Edagawa, Y. Yoshida, and H. Wakabayashi, "Long-haul coherent optical fiber communication systems using optical amplifiers," J. Lightwave Technol. 9(2), 251–260 (1991).

[9] A. H. Gnauck and P. J. Winzer, "Optical phase-shift-keyed transmission," J. Lightwave Technol. 23(1), 115–130 (2005).

[10] M. Nakazawa, "Optical Quadrature Amplitude Modulation (QAM) with Coherent Detection up to 128 States," in Proc. Optical Fiber Communication Conference (OFC'09), paper OThG1 (2009).

[11] K. Roberts, D. Beckett, D. Boertjes, J. Berthold, and C. Laperle, "100G and beyond with digital coherent signal processing," IEEE Commun. Mag. 48(7), 62–69 (2010).

[12] Y. Miyamoto, A. Hirano, K. Yonenaga, A. Sano, H. Toba, K. Murata, and O. Mitomi, "320 Gbit/s (8×40 Gbit/s) WDM transmission over 367 km with 120 km repeater spacing using carrier-suppressed return-to-zero format," Electronics Letters 35(23), 2041–2042 (1999).

[13] T. Hirooka, P. Ruan, P. Guan, and M. Nakazawa, "Highly dispersion-tolerant 160 Gbaud optical Nyquist pulse TDM transmission over 525 km," Opt. Express 20, 15001–15007 (2012).

[14] M. Nakazawa, T. Hirooka, P. Ruan, and P. Guan, "Ultrahigh-speed "orthogonal" TDM transmission with an optical Nyquist pulse train," Opt. Express 20, 1129–1140 (2012).

[15] I. Neokosmidis, T. Kamalakis, and T. Sphicopoulos, "Nonlinearity tolerance of optical modulation formats in nonzero dispersion fibers," IEEE Photon. Technol. Lett. 17(12), 2760–2762 (2005).

[16] J. Salz, "Coherent lightwave communications." AT&T technical journal 64(10), 2153–2207 (1985).

[17] T. A. Birks, D. Mogilevtsev, J. C. Knight, and P. S. J. Russell, "Dispersion compensation using single-material fibers," IEEE Photon. Technol. Lett. 11(6), 674–676 (1999).

[18] S. Savory, G. Gavioli, R. Killey, and P. Bayvel, "Electronic compensation of chromatic dispersion using a digital coherent receiver," Opt. Express 15(5), 2120–2126 (2007).

[19] X. Zhou, L. Nelson, P. Magill, B. Zhu, and D. Peckham, "8x450-Gb/s, 50-GHz-spaced, PDM-32QAM transmission over 400km and one 50GHz-grid ROADM," in Optical Fiber Communications Conference (OFC'11), paper PDPB3 (2011).

[20] S. Bigo, Y. Frignac, G. Charlet, W. Idler, S. Borne, H. Gross, R. Dischler, W. Poehlmann, P. Tran, C. Simonneau, D. Bayart, G. Veith, A. Jourdan, and J. Hamaide, "10.2Tbit/s (256x42.7Gbit/s PDM/WDM) transmission over 100km TeraLight fiber with 1.28bit/s/Hz spectral efficiency," in Proc. Optical Fiber Communication Conference (OFC'01), paper PD25 (2001).

[21] H. Mulvad, M. Galili, L. Oxenløwe, H. Hu, A. Clausen, J. Jensen, C. Peucheret, and P. Jeppesen, "Demonstration of 5.1 Tbit/s data capacity on a single-wavelength channel," Opt. Express 18(2), 1438–1443 (2010).

[22] C. Jin-Xing, C. R. Davidson, A. Lucero, Z. Hongbin, D. G. Foursa, O. V. Sinkin, W. W. Patterson, A. N. Pilipetskii, G. Mohs, and N. S. Bergano, "20 Tbit/s Transmission Over 6860 km With Sub-Nyquist Channel Spacing," J. Ligthwave Technol. 30(4), 651–657 (2012).

[23] W. Shieh, H. Bao, and Y. Tang, "Coherent optical OFDM: theory and design," Opt. Express **16**(2), 841–859 (2008).

[24] G. Bosco, V. Curri, A. Carena, P. Poggiolini, and F. Forghieri, "On the Performance of Nyquist-WDM Terabit Superchannels Based on PM-BPSK, PM-QPSK, PM-8QAM or PM-16QAM Subcarriers," J. Lightwave Technol. **29**, 53–61 (2011).

[25] P. J. Winzer, "Energy-Efficient Optical Transport Capacity Scaling Through Spatial Multiplexing," IEEE Photon. Technol. Lett. 23(13), 851–853 (2011).

[26] B. Puttnam, J. Delgado-Mendinueta, J. Sakaguchi, R. Luís, W. Klaus, Y. Awaji, N. Wada, A. Kanno, and T. Kawanishi, "105Tb/s Transmission System Using Low-cost, MHz Linewidth DFB Lasers Enabled by Self-Homodyne Coherent Detection and a 19-Core Fiber," in Proc. Optical Fiber Communication Conference (OFC'13), paper OW1I.1 (2013).

[27] R. Ryf, S. Randel, N. Fontaine, M. Montoliu, E. Burrows, S. Chandrasekhar, A. Gnauck, C. Xie, R. Essiambre, P. Winzer, R. Delbue, P. Pupalaikis, A. Sureka, Y. Sun, L. Gruner-Nielsen, R. Jensen, and R. Lingle, "32-bit/s/Hz Spectral Efficiency WDM Transmission over 177-km Few-Mode Fiber," in Proc. Optical Fiber Communication Conference (OFC'13), paper PDP5A.1 (2013).

[28] E. Ip, M. Li, Y. Huang, A. Tanaka, E. Mateo, W. Wood, J. Hu, Y. Yano, and K. Koreshkov, "146λ×6×19-Gbaud Wavelength- and Mode-Division Multiplexed Transmission over 10×50-km Spans of Few-Mode Fiber with a Gain-Equalized Few-Mode EDFA," in Proc. Optical Fiber Communication Conference (OFC'13), paper PDP5A.2 (2013).

[29] P. Winzer and G. Foschini, "MIMO capacities and outage probabilities in spatially multiplexed optical transport systems," Opt. Express 19(17), 16680–16696 (2011).

[30] Q. Dayou, H. Ming-Fang, I. Ezra, H. Yue-Kai, S. Yin, H. Junqiang, and W. Ting, "101.7-Tb/s (370×294-Gb/s) PDM-128QAM-OFDM transmission over 3×55-km SSMF using pilot-based phase noise mitigation," in Proc. Optical Fiber Communication Conference (OFC'11), paper PDPB5 (2011).

[31] D. Hillerkuss, R. Schmogrow, M. Meyer, S. Wolf, M. Jordan, P. Kleinow, N. Lindenmann, P. Schindler, A. Melikyan, X. Yang, S. Ben-Ezra, B. Nebendahl, M. Dreschmann, J. Meyer, F. Parmigiani, P. Petropoulos, B. Resan, A. Oehler, K. Weingarten, L. Altenhain, T. Ellermeyer, M. Moeller, M. Huebner, J. Becker, C. Koos, W. Freude, and J. Leuthold, "Single-Laser 32.5 Tbit/s Nyquist WDM Transmission," J. Opt. Commun. Netw. 4(10), 715–723 (2012).

[32] Y. Benlachtar, P. M. Watts, R. Bouziane, P. Milder, D. Rangaraj, A. Cartolano, R. Koutsoyannis, J. C. Hoe, M. Püschel, M. Glick, and R. I. Killey, “Generation of optical OFDM signals using 21.4 GS/s real time digital signal processing,” Opt. Express 17(20), 17658–17668 (2009).

[33] X. Q. Jin, R. P. Giddings, E. Hugues-Salas, and J. M. Tang, “Real-time demonstration of 128-QAM-encoded optical OFDM transmission with a 5.25bit/s/Hz spectral efficiency in simple IMDD systems utilizing directly modulated DFB lasers,” Opt. Express 17(22), 20484–20493 (2009).

[34] F. Buchali, R. Dischler, A. Klekamp, M. Bernhard, and Y. Ma, “Statistical transmission experiments using a real-time 12.1 Gb/s OFDM transmitter,” in Proc. Optical Fiber Communication Conference (OFC’10), paper OMS3 (2010).

[35] N. Kaneda, Q. Yang, X. Liu, S. Chandrasekhar, W. Shieh, and Y. Chen, “Real-time 2.5 GS/s coherent optical receiver for 53.3-Gb/s sub-banded OFDM,” J. Lightwave Technol. 28(4), 494–501 (2010).

[36] D. Qian, T. Kwok, N. Cvijetic, J. Hu, and T. Wang, “41.25 Gb/s real-time OFDM receiver for variable rate WDM-OFDMA-PON transmission,” in Proc. Optical Fiber Communication Conference (OFC’10), paper PDPD9 (2010).

[37] B. Inan, O. Karakaya, P. Kainzmaier, S. Adhikari, S. Calabro, V. Sleiffer, N. Hanik, and S. Jansen, “Realization of a 23.9 Gb/s real time optical-OFDM transmitter with a 1024 Point IFFT,” in Proc. Optical Fiber Communication Conference (OFC’11), paper OMS2 (2011).

[38] B. Inan, S. Adhikari, O. Karakaya, P. Kainzmaier, M. Mocker, H. von Kirchbauer, N. Hanik, and S. Jansen, "Real-time 93.8-Gb/s polarization-multiplexed OFDM transmitter with 1024-point IFFT," Opt. Express 19(26), B64–B68 (2011).

[39] R. Schmogrow, M. Winter, D. Hillerkuss, B. Nebendahl, S. Ben-Ezra, J. Meyer, M. Dreschmann, M. Huebner, J. Becker, C. Koos, W. Freude, and J. Leuthold, “Real-time OFDM transmitter beyond 100 Gbit/s,” Opt. Express 19(13), 12740–12749 (2011).

[40] R. Schmogrow, M. Winter, M. Meyer, D. Hillerkuss, B. Nebendahl, J. Meyer, M. Dreschmann, M. Huebner, J. Becker, C. Koos, W. Freude, and J. Leuthold, “Real-Time Nyquist Pulse Modulation Transmitter Generating Rectangular Shaped Spectra of 112 Gbit/s 16QAM Signals,” in OSA Topical Meeting: Signal Processing in Photonic Communications (SPPCom’11), paper SPMA5 (2011).

[41] R. Schmogrow, M. Meyer, S. Wolf, B. Nebendahl, D. Hillerkuss, B. Baeuerle, M. Dreschmann, J. Meyer, M. Huebner, J. Becker, C. Koos, W. Freude, and J. Leuthold, “150 Gbit/s Real-Time Nyquist Pulse Transmission Over 150 km

SSMF Enhanced by DSP with Dynamic Precision," in Proc. Optical Fiber Communication Conference (OFC'12), paper OM2A.6 (2012).

[42] R. Schmogrow, R. Bouziane, D. Hillerkuss, P. Milder, R. Koutsoyannis, R. Killey, Y. Benlachtar, P. Watts, P. Bayvel, C. Koos, W. Freude, and J. Leuthold, "85.4 Gbit/s Real-Time OFDM Signal Generation with Transmission over 400 km and Preamble-less Reception," in Proc. Conf. on Lasers and Electro-Optics (CLEO'12), paper CTh1H.4 (2012).

[43] R. Bouziane, R. Schmogrow, D. Hillerkuss, P. Milder, C. Koos, W. Freude, J. Leuthold, P. Bayvel, and R. Killey, "Generation and transmission of 85.4 Gb/s real-time 16QAM coherent optical OFDM signals over 400 km SSMF with preamble-less reception," Opt. Express 20(19), 21612–21617 (2012).

[44] R. Schmogrow, M. Meyer, P. Schindler, A. Josten, S. Ben-Ezra, C. Koos, W. Freude, and J. Leuthold, "252 Gbit/s Real-Time Nyquist Pulse Generation by Reducing the Oversampling Factor to 1.33," in Proc. Optical Fiber Communication Conference (OFC'13), paper OTu2I.1 (2013).

[45] S. K. Kurtz and F. N. H. Robinson, "A Physical Model of the Electro-Optic Effect," Applied Physics Letters 10(2), 62–65 (1967).

[46] N. F. Borrelli, "Electro-optic Effect in Transparent Niobate Glass—Ceramic Systems," Journal of Applied Physics 38(11), 4243–4247 (1967).

[47] A. Katz, W. Jemison, M. Kubak, and J. Dragone, "Improved radio over fiber performance using predistortion linearization," IEEE MTT-S International Microwave Symposium Digest 2, 1403–1406 (2003).

[48] I. Fatadin, S. J. Savory, and D. Ives, "Compensation of Quadrature Imbalance in an Optical QPSK Coherent Receiver," IEEE Photon. Technol. Lett. 20(20), 1733–1735 (2008).

[49] E. Snitzer, "Cylindrical Dielectric Waveguide Modes," J. Opt. Soc. Am. 51, 491–498 (1961).

[50] F. Parmigiani, C. Finot, K. Mukasa, M. Ibsen, M. Roelens, P. Petropoulos, and D. Richardson, "Ultra-flat SPM-broadened spectra in a highly nonlinear fiber using parabolic pulses formed in a fiber Bragg grating," Opt. Express 14, 7617–7622 (2006).

[51] P. Russell, "Photonic-Crystal Fibers," J. Lightwave Technol. 24, 4729–4749 (2006).

[52] S. Randel, R. Ryf, A. Sierra, P. Winzer, A. Gnauck, C. Bolle, R. Essiambre, D. Peckham, A. McCurdy, and R. Lingle, "6×56-Gb/s mode-division multiplexed

transmission over 33-km few-mode fiber enabled by 6×6 MIMO equalization," Opt. Express 19, 16697–16707 (2011).

[53] T. Hayashi, T. Taru, O. Shimakawa, T. Sasaki, and E. Sasaoka, "Design and fabrication of ultra-low crosstalk and low-loss multi-core fiber," Opt. Express 19, 16576–16592 (2011).

[54] C. Herard and A. Lacourt, "New multiplexing technique using polarization of light," Appl. Opt., vol. 30, no. 2, pp. 222–231, 1991.

[55] P. K. A. Wai and C. R. Menyak, "Polarization mode dispersion, decorrelation, and diffusion in optical fibers with randomly varying birefringence," J. Lightwave Technol. 14(2), 148–157 (1996).

[56] A. Mecozzi, M. Shtaif, "The statistics of polarization-dependent loss in optical communication systems," IEEE Photon. Technol. Lett. 14(3), 313–315 (2002).

[57] R. Olshansky, "Noise figure for erbium-doped optical fibre amplifiers," Electron. Lett. 24(22), 1363–1365, (1988).

[58] A. Tzanakaki, and M. J. O'Mahony, "Analysis of tunable wavelength converters based on cross-gain modulation in semiconductor optical amplifiers operating in the counter propagating mode,", IEE Proceedings Optoelectronics, vol. 147, 49–55, 2000.

[59] I. P. Kaminow, T. Li, and A. E. Wilner (eds.), Optical Fiber Telecommunications, B: Systems and Networks, (Academic Press, USA, 2008).

[60] S. Okamoto, K. Toyoda, T. Omiya, K. Kasai, M. Yoshida, and M. Nakazawa, "512 QAM (54 Gbit/s) coherent optical transmission over 150 km with an optical bandwidth of 4.1 GHz," in 36th European Conference and Exhibition on Optical Communication (ECOC'10), paper PD2.3.

[61] H. Nyquist, "Certain topics in telegraph transmission theory," Trans. Am. Inst. Electr. Eng., vol. 47, 617–644, 1928.

[62] A. Jerri, "The Shannon sampling theorem—Its various extensions and applications: A tutorial review," Proc. IEEE, vol. 65, no. 11, pp. 1565–1596, 1977.

[63] S. Smith, The Scientist & Engineer's Guide to Digital Signal Processing, (California Technical Pub., USA, 1997), p. 45.

[64] E. T. Whittaker, "On the functions which are represented by the expansions of the interpolation theory," Proc. R. Soc. Edinburgh, vol. 35, pp. 181–194, 1915.

[65] R. Waegemans, S. Herbst, L. Holbein, P. Watts, P. Bayvel, C. Fürst, and R. I. Killey, "10.7 Gb/s electronic predistortion transmitter using commercial FPGAs

and D/A converters implementing real-time DSP for chromatic dispersion and SPM compensation," Opt. Express 17, 8630–8640 (2009).

[66] H. Sun, K. Wu, and K. Roberts, "Real-time measurements of a 40 Gb/s coherent system," Opt. Express 16(2), 873–879 (2008).

[67] R. W. Tkach, R. M. Derosier, A. H. Gnauck, A. M. Vengsarkar, D. W. Peckham, J. L. Zyskind, J. W. Sulhoff, and A. R. Chraplyvy, "Transmission of eight 20-Gb/s channels over 232 km of conventional single-mode fiber," IEEE Photon. Technol. Lett. 7(11), 1369–1371 (1995).

[68] S. J. Savory, "Digital filters for coherent optical receivers," Opt. Express 16(2), 804–817 (2008).

[69] R. Noe, S. Hinz, D. Sandel, and F. Wust, "Crosstalk detection schemes for polarization division multiplex transmission," J. Lightwave Technol. 19(10), 1469–1475 (2001).

[70] C. R. Johnson, Jr., P. Schniter, T. J. Endres, J. D. Behm, D. R. Brown, and R. A. Casas, "Blind equalization using the constant modulus criterion: a review," P. IEEE 86(10), 1927–1950 (1998).

[71] B. Szafraniec, B. Nebendahl, and T. Marshall, "Polarization demultiplexing in Stokes space," Opt. Express 18(17), 17928–17939 (2010).

[72] M. Morsy-Osman, M. Chagnon, Q. Zhuge, X. Xu, M. Mousa-Pasandi, Z. El-Sahn, and D. Plant, "Ultrafast and low overhead training symbol based channel estimation in coherent M-QAM single-carrier transmission systems," Opt. Express 20(26), B171–B180 (2012).

[73] R. Schmogrow, D. Hillerkuss, M. Dreschmann, M. Huebner, M. Winter, J. Meyer, B. Nebendahl, C. Koos, J. Becker, W. Freude, and J. Leuthold, "Real-time software-defined multiformat transmitter generating 64QAM at 28 GBd", IEEE Photon. Technol. Lett. 22(21), 1601–1603 (2010).

[74] S. D. Personick, "Receiver design for digital fiber optic communication systems, I," Bell Syst. Tech. J. 52, 843–874 (1973).

[75] R. Noé, T. Pfau, M. El-Darawy, and S. Hoffmann, "Electronic Polarization Control Algorithms for Coherent Optical Transmission," IEEE Journal of Selected Topics in Quantum Electronics 16(5), 1193–1200 (2010).

[76] D. O. North, "An analysis of the factors which determine signal/noise discrimination in pulsed carrier systems," RCA Labs, Princeton, NJ, Rep. PTR-6C (1943).

[77] J. G. Proakis, Digital Communications, 5th revised edition (McGraw-Hill Higher Education, New York, USA, 2008).

[78] M. Sjödin, P. Johannisson, H. Wymeersch, P. Andrekson, and M. Karlsson, "Comparison of polarization-switched QPSK and polarization-multiplexed QPSK at 30 Gbit/s," Opt. Express 19, 7839–7846 (2011).

[79] R. Schmogrow, M. Winter, M. Meyer, D. Hillerkuss, S. Wolf, B. Baeuerle, A. Ludwig, B. Nebendahl, S. Ben-Ezra, J. Meyer, M. Dreschmann, M. Huebner, J. Becker, C. Koos, W. Freude, and J. Leuthold, "Real-time Nyquist pulse generation beyond 100 Gbit/s and its relation to OFDM," Opt. Express 20, 317–337 (2012).

[80] G. Clark, S. Mitra, and S. Parker, "Block implementation of adaptive digital filters," IEEE T. Circuits Syst. 28, 584–592 (1981).

[81] R. Schmogrow, B. Nebendahl, M. Winter, A. Josten, D. Hillerkuss, S. Koenig, J. Meyer, M. Dreschmann, M. Huebner, C. Koos, J. Becker, W. Freude, and J. Leuthold, "Error vector magnitude as a performance measure for advanced modulation formats," IEEE Photon. Technol. Lett. 24(1), 61–63 (2012).

[82] R. Schmogrow, B. Nebendahl, M. Winter, A. Josten, D. Hillerkuss, S. Koenig, J. Meyer, M. Dreschmann, M. Huebner, C. Koos, J. Becker, W. Freude, and J. Leuthold, "Corrections to "Error Vector Magnitude as a Performance Measure for Advanced Modulation Formats" [Jan 1, 2012 61-63]," IEEE Photon. Technol. Lett. 24(23), 2198 (2012).

[83] R. Schmogrow, D. Hillerkuss, S. Wolf, B. Bäuerle, M. Winter, P. Kleinow, B. Nebendahl, T. Dippon, P. Schindler, C. Koos, W. Freude, and J. Leuthold, "512QAM Nyquist sinc-pulse transmission at 54 Gbit/s in an optical bandwidth of 3 GHz," Opt. Express 20(6), 6439–6447 (2012).

[84] J. Leuthold and W. Freude, "Optical OFDM and Nyquist multiplexing," in Optical Fiber Telecommunications Volume VIB: Systems and Networks (Optics and Photonics), Edited by I. Kaminow, T. Li, A.E. Willner, Academic Press (2013).

[85] F. M. Gardner, "A BPSK/QPSK timing-error detector for sampled receivers," IEEE Trans. Comm. 34(5), 423–429 (1986)

[86] H. Meyr, M. Moeneclaey, and S.A. Fechtel, "Digital communication receivers: Synchronization, channel estimation, and signal processing," New York: Wiley (1997).

[87] N. D'Andrea and U. Mengali, "A simulation study of clock recovery in QPSK and 9QPRS systems," IEEE Trans. Comm. 36(5), 605–612 (1985).

[88] M. Örder and H. Meyr, "Digital filter and square timing recovery," IEEE Trans. Commun. 36(5), 605–612 (1988).

[89] M. Yan, Z. Tao, L. Dou, L. Li, Y. Zhao, T. Hoshida, and J. Rasmussen, "Digital Clock Recovery Algorithm for Nyquist Signal," in Proc. Optical Fiber Communication Conference (OFC'13), paper OTu2I.7 (2013).

[90] A. Viterbi, "Nonlinear estimation of PSK-modulated carrier phase with application to burst digital transmission," IEEE Transactions on Information Theory 29(4), 543–551 (1983).

[91] T. M. Schmidl and D. C. Cox, "Robust frequency and timing synchronization for OFDM," IEEE Trans. Commun. 45, 1613–1621 (1997).

[92] J.E. Volder, "The CORDIC trigonometric computing technique," IRE Transactions on Electronic Computers, 330–334 (1959).

[93] C.R. Johnson, Jr., P. Schniter, T.J. Endres, J.D. Behm, D.R. Brown, and R.A. Casas, "Blind equalization using the constant modulus criterion: A review," Proc. IEEE 86(10), 1927–1950 (1998).

[94] K. Hoon and A. H. Gnauck, "Experimental investigation of the performance limitation of DPSK systems due to nonlinear phase noise," IEEE Photon. Technol. Lett. 15(2), 320–322 (2003).

[95] S. Haykin, Adaptive Filter Theory, (Prentice Hall Information and Systems Sciences Series, USA, 1996), 3rd revised edition, Chapters 5 and 9.

[96] B. Widrow and E. Hoff, "Adaptive switching circuits," Stanford (1960). Available: http://isl-www.stanford.edu/~widrow/papers/c1960adaptiveswitching.pdf

[97] G. Clark, S. Mitra, and S. Parker, "Block implementation of adaptive digital filters," IEEE Transactions on Acoustics, Speech and Signal Processing, 29(3), 744–752 (1981).

[98] "Virtex™ 2.5 V Field Programmable Gate Arrays," Xilinx, Inc., San Jose, CA, USA, product specification, April 2001.

[99] "Virtex-II Platform FPGAs: Complete Data Sheet," Xilinx, Inc., San Jose, CA, USA, product specification, November 2007.

[100] "Virtex-II Pro and Virtex-II Pro X Platform FPGAs: Complete Data Sheet," Xilinx, Inc., San Jose, CA, USA, product specification, June 2011.

[101] "Virtex-4 Family Overview," Xilinx, Inc., San Jose, CA, USA, product specification, August 2010.

[102] "Virtex-5 Family Overview," Xilinx, Inc., San Jose, CA, USA, product specification, February 2009.

[103] "Virtex-6 Family Overview," Xilinx, Inc., San Jose, CA, USA, product specification, January 2012.

[104] "7 Series FPGAs Overview," Xilinx, Inc., San Jose, CA, USA, advance product specification, May 2012.

[105] "Programming FPGAs with LabVIEW Statecharts," National Instruments, Austin, TX, USA, white paper, March 2011. [Online]. Available: http://www.ni.com/white-paper/6221/en

[106] "HDL Coder," Mathworks, Natick, MA, USA, datasheet, 2012. [Online]. Available: http://www.mathworks.com/products/hdl-coder

[107] "Vivado Design Suite," Xilinx, Inc., San Jose, CA, USA, product brief, 2012. [Online]. Available: http://www.xilinx.com/publications/prod_mktg/vivado/Vivado_Product_Brief.pdf

[108] W. R. Sutherland, "On-Line Graphical Specification of Computer Procedures," Mass. Iinst. of Tech. (MIT) Lexington, Lincoln Lab (1966).

[109] A. H. Gnauck, G. Raybon, S. Chandrasekhar, J. Leuthold, C. Doerr, L. Stulz, A. Agarwal, S. Banerjee, D. Grosz, S. Hunsche, A. Kung, A. Marhelyuk, D. Maywar, M. Movassaghi, X. Liu, C. Xu, X. Wei, and D. M. Gill, "2.5 Tb/s (64'42.7 Gb/s) transmission over 40'100 km NZDSF using RZ-DPSK format and all-Raman-amplified spans", in Proc. Optical Fiber Communication Conference (OFC'02), paper PDPFC2 (2002).

[110] M. Nakazawa, S. Okamoto, T. Omiya, K. Kasai, and M. Yoshida, "256-QAM (64 Gb/s) coherent optical transmission over 160 km with an optical bandwidth of 5.4 GHz," IEEE Photon. Tech. Lett. 22(3), 185–187 (2010).

[111] A. H. Gnauck, P. J. Winzer, S. Chandrasekhar, X. Liu, B. Zhu, and D. W. Peckham, "10 × 224-Gb/s WDM transmission of 28-Gbaud PDM 16-QAM on a 50-GHz grid over 1,200 km of fiber," in Proc. Optical Fiber Communication Conference (OFC'10), paper PDPB8 (2010).

[112] M. Secondini, E. Forestieri, and F. Cavaliere, "Novel optical modulation scheme for 16-QAM format with quadrant differential encoding," in Photonics in Switching (PS'09), (2009).

[113] A. Chiba, T. Sakamoto, T. Kawanishi, K. Higuma, M. Sudo, and J. Ichikawa, "16-level quadrature amplitude modulation by monolithic quad-parallel mach-zehnder optical modulator," Electron. Lett. 46(3), 227–228 (2010).

[114] D. Hillerkuss, R. Schmogrow, M. Huebner, M. Winter, B. Nebendahl, J. Becker, W. Freude, and J. Leuthold, "Software-defined multi-format transmitter with real-

time signal processing for up to 160 Gbit/s", in OSA Topical Meeting: Signal Processing in Photonic Communications (SPPCom'10), paper SPTuC4 (2010).

[115] R.A. Shafik, M. S. Rahman, and A. H. M. R. Islam, "On the extended relationships among EVM, BER and SNR as performance metrics", 4th International Conference on Electrical and Computer Engineering (ICECE'06), 408–411 (2006).

[116] D. Hillerkuss, R. Schmogrow, T. Schellinger, M. Jordan, M. Winter, G. Huber, T. Vallaitis, R. Bonk, P. Kleinow, F. Frey, M. Roeger, S. Koenig, A. Ludwig, A. Marculescu, J. Li, M. Hoh, M. Dreschmann, J. Meyer, S. Ben Ezra, N. Narkiss, B. Nebendahl, F. Parmigiani, P. Petropoulos, B. Resan, A. Oehler, K. Weingarten, T. Ellermeyer, J. Lutz, M. Moeller, M. Huebner, J. Becker, C. Koos, W. Freude, and J. Leuthold, "26 Tbit s−1 line-rate super-channel transmission utilizing all-optical fast Fourier transform processing," Nat. Photonics 5, 364–371 (2011).

[117] X. Liu, S. Chandrasekhar, B. Zhu, P. Winzer, A. Gnauck, and D. Peckham, "448-Gb/s reduced-guard-interval CO-OFDM transmission over 2000 km of ultra-large-area fiber and five 80-GHz-grid ROADMs," J. Lightw. Technol. 29, 483–490 (2011).

[118] M. Nakazawa, S. Okamoto, T. Omiya, K. Kasai, and M. Yoshida, "256 QAM (64 Gbit/s) coherent optical transmission over 160 km with an optical bandwidth of 5.4 GHz," in Proc. Optical Fiber Communications Conference (OFC'10), paper OMJ5 (2010).

[119] R. Essiambre, G. Kramer, P. Winzer, G. Foschini, and B. Goebel, "Capacity limits of optical fiber networks," J. Lightw. Technol. 28, 662–701 (2010)

[120] A. Ellis, J. Zhao, and D. Cotter, "Approaching the non-linear Shannon limit," J. Lightw. Technol. 28, 423–433 (2010).

[121] Z. Dong, J. Yu, H. Chien, N. Chi, L. Chen, and G. Chang, "Ultra-dense WDM-PON delivering carrier-centralized Nyquist WDM uplink with digital coherent detection," Opt. Express 19, 11100–11105 (2011).

[122] L. Lo Presti and M. Mondin, "Design of optimal FIR raised-cosine filters," Electron. Lett. 25, 467–468 (1989).

[123] E. Sun, B. Tian, Y. Wang, and K. Yi, "Quasi-orthogonal time division multiplexing and its applications under Rayleigh fading channels," in 21st International Conference on Advanced Information Networking and Applications Workshops, 172–176 (2007)

[124] M. Abramowitz and I. A. Stegun (Eds.) "Exponential Integral and Related Functions," in Handbook of Mathematical Functions, New York: Dover (1972). Formula 5.2.1 on Page 231.

[125] I. S. Gradstein and I. M. Ryshik (Eds.) Tables of Series, Products, and Integrals, Vol. 1 and 2. Thun: Harri Deutsch (1981).

[126] P. Milder, R. Bouziane, R. Koutsoyannis, C. Berger, Y. Benlachtar, R. Killey, M. Glick, and J. Hoe, "Design and Simulation of 25 Gb/s Optical OFDM Transceiver ASICs," in 37th European Conference and Exposition on Optical Communications (ECOC'11), paper We.9.A.5 (2011).

[127] D. Rafique and A. Ellis, "Nonlinear penalties in long-haul optical networks employing dynamic transponders," Opt. Express 19, 9044–9049 (2011).

[128] M. Yoshida, T. Omiya, K. Kasai, and M. Nakazawa, "Real-time FPGA-based coherent optical receiver for 1 Gsymbol/s, 64 QAM transmission," in Optical Fiber Communication Conference (OFC'11), paper OTuN3 (2011).

[129] R. Schmogrow, R. Bouziane, M. Meyer, P. Milder, P. Schindler, R. Killey, P. Bayvel, C. Koos, W. Freude, and J. Leuthold, "Real-time OFDM or Nyquist pulse generation – which performs better with limited resources?," Opt. Express 20(26), B543–B551 (2012).

[130] R. Schmogrow, S. Ben-Ezra, P.C. Schindler, B. Nebendahl, C. Koos, W. Freude, and J. Leuthold, "Pulse-shaping with digital, electrical, and optical filter – a comparison," J. Lightw. Technol. 31(15), 2570–2577 (2013).

[131] A. Nespola, S. Straullu, G. Bosco, A. Carena, Yanchao Jiang, P. Poggiolini, F. Forghieri, Y. Yamamoto, M. Hirano, T. Sasaki, J. Bauwelinck, and K. Verheyen, "1306-km 20×124.8-Gb/s PM-64QAM transmission over PSCF with net SEDP 11,300 (bkm)/s/Hz using 1.15 samp/symb DAC," in 39th European Conference and Exhibition on Optical Communication (ECOC'13), paper Th.2.D.1 (2013).

[132] T. Mizuochi, "Recent progress in forward error correction and its interplay with transmission impairments," IEEE J. Sel. Topics Quantum Electron. 12, 544–554 (2006).

[133] N. Cvijetic, "Optical OFDM for next-generation PON," in OSA Topical Meeting: Signal Processing in Photonic Communications (SPPCom'10), paper SPTuB6 (2010).

[134] X. Liu, S. Chandrasekhar, B. Zhu, and D. Peckham, "Efficient digital coherent detection of a 1.2-Tb/s 24-carrier no-guard-interval CO-OFDM signal by simultaneously detecting multiple carriers per sampling," in Proc. Optical Fiber Communications Conference (OFC'10), paper OWO2 (2010).

[135] R. Schmogrow, M. Winter, B. Nebendahl, D. Hillerkuss, J. Meyer, M. Dreschmann, M. Huebner, J. Becker, C. Koos, W. Freude, and J. Leuthold, "101.5 Gbit/s real-time OFDM transmitter with 16QAM modulated subcarriers," in Proc. Optical Fiber Communications Conference (OFC'11), paper OWE5 (2011).

[136] Q. Yang, N. Kaneda, X. Liu, S. Chandrasekhar, W. Shieh, and Y. Chen, "Towards Real-Time Implementation of Optical OFDM Transmission," in Proc. Optical Fiber Communications Conference (OFC'10), paper OMS6 (2010).

[137] S. G. Johnson, M. Frigo, "A modified split-radix FFT with fewer arithmetic operations," IEEE Trans. Signal Process. 55(1), 111–119 (2007).

[138] N. Cvijetic, "OFDM for Next Generation Optical Access Networks," J. Lightw. Technol. 30(4), 384–398 (2012).

[139] X.Q. Jin, E. Hugues-Salas, R.P. Giddings, J. L. Wei, J. Groenewald, and J. M. Tang, "First Real-time Experimental Demonstrations of 11.25Gb/s Optical OFDMA PONs with Adaptive Dynamic Bandwidth Allocation," Opt. Express, 19(21), 20557–20570 (2011).

[140] Y. Benlachtar, R. Bouziane, R. I. Killey, C. R. Berger, P. Milder, R. Koutsoyannis, J. C. Hoe, M. Püschel, and M. Glick, "Optical OFDM for the data center," in Proc. of International Conference on Transparent Optical Networks (ICTON'10), (2010).

[141] Y. Benlachtar, P. M. Watts, R. Bouziane, P. Milder, R. Koutsoyannis, J. C. Hoe, M. Püschel, M. Glick, and R. I. Killey, "Real-Time Digital Signal Processing for the Generation of Optical Orthogonal Frequency-Division-Multiplexed Signals," IEEE Journal of Selected Topics in Quantum Electronics 16(5), 1235–1244 (2010).

[142] P. A. Milder, F. Franchetti, J. C. Hoe, and M. Püschel, "Computer Generation of Hardware for Linear Digital Signal Processing Transforms," ACM Transactions on Design Automation of Electronic Systems 17(2), Article 15 (2012).

[143] UG198 Virtex-5 FPGA RocketIO GTX Transceiver User Guide (2009), Xilinx Inc., Available: http://www.xilinx.com/support/documentation/user_guides/ug198.pdf

[144] C. Simin, M. Yiran, and W. Shieh, "Multiband real-time coherent optical OFDM reception up to 110 Gb/s with 600-km transmission," IEEE Photon. J. 2(3), 454–459 (2010).

[145] R. Schmogrow, R. Bouziane, M. Meyer, P. A. Milder, P. C. Schindler, P. Bayvel, R. I. Killey, W. Freude, and J. Leuthold, "Digital Nyquist-WDM and OFDM Signal Generation: Spectral Efficiency Versus DSP Complexity," in European

Conference and Exhibition on Optical Communication (ECOC'12), paper Mo.2.A.4 (2012).

[146] B. Spinnler, "Equalizer design and complexity for digital coherent receivers," IEEE J Quantum Elect. 16(5), 1180–1192 (2010).

[147] G. Bosco, A. Carena, V. Curri, P. Poggiolini, and F. Forghieri, "Performance limits of Nyquist-WDM and CO-OFDM in high-speed PM-QPSK systems," IEEE Photon. Technol. Lett. 22, 1129–1131 (2010).

[148] N. Bergano, "Wavelength division multiplexing in long-haul transmission systems," in Proc. Frontiers in Optics, paper FMF4 (2004).

[149] C. A. Brackett, "Dense wavelength division multiplexing networks: principles and applications," J. Sel. Area. Comm. 8, 948–964 (1990).

[150] B. Zhu, L. E. Nelson, S. Stulz, A. H. Gnauck, C. Doerr, J. Leuthold, L. Gruner-Nielsen, M. O. Pedersen, J. Kim, and R. L. Lingle Jr., "High spectral density long-haul 40-Gb/s transmission using CSRZ-DPSK format," J. Lightwave Technol. 22, 208–214 (2004).

[151] J. Cartledge, J. Downie, J. Hurley, A. Karar, Y. Jiang, and K. Roberts, "Pulse shaping for 112 Gbit/s polarization multiplexed 16-QAM signals using a 21 GSa/s DAC," Opt. Express 19, B628–B635 (2011).

[152] R. Schmogrow, S. Wolf, B. Baeuerle, D. Hillerkuss, B. Nebendahl, C. Koos, W. Freude, and J. Leuthold, "Nyquist Frequency Division Multiplexing for Optical Communications," in Proc. Conf. on Lasers and Electro-Optics (CLEO'12), paper CTh1H.2 (2012).

[153] M. S. O'Sillivan, K. Roberts, and C. Bontu, "Electronic dispersion compensation techniques for optical communication systems," in 31st European Conference and Exhibition on Optical Communication (ECOC'05), paper Tu3.2.1 (2005).

[154] C. Xia and W. Rosenkranz, "Electrical dispersion compensation for different modulation formats with optical filtering," in Proc. Optical Fiber Communication Conference (OFC'06), paper OWR2 (2006).

[155] C. E. Shannon, "A mathematical theory of communication," Bell Syst. Tech. J. 27, 623–656 (1948).

[156] Telecommunication standardization sector of International telecommunication union (ITU-T), Transmission system and media, digital system and networks G.975.1.

[157] T. Kremp and W. Freude, "Fast Split-Step Wavelet Collocation Method for WDM System Parameter Optimization," J. Lightwave Technol. 23, 1491–1502 (2005).

[158] K.-P. Ho, "Error probability of DPSK signals with cross-phase modulation induced nonlinear phase noise," IEEE J. Quantum Electron. 10, 421–427 (2004).

[159] S. Adhikari, B. Inan, O. Karakaya, W. Rosenkranz, and S. L. Jansen, "FFT optimization for practical OFDM implementations," in 37th European Conference and Exhibition on Optical Communication (ECOC'11), paper We.P1.25 (2011).

[160] S. H. Muller-Weinfurtner, "Optimum Nyquist windowing in OFDM receivers," IEEE Transactions on Communications 49(3), 417–420 (2001).

[161] K. Christodoulopoulos, I. Tomkos, and E. Varvarigos, "Dynamic Bandwidth Allocation in Flexible OFDM-based Networks," in Proc. Optical Fiber Communication Conference (OFC'11), paper OTuI5 (2011).

[162] M. Huang, D. Qian, and N. Cvijetic, "A Novel Symmetric Lightwave Centralized WDM-OFDM-PON Architecture with OFDM-Remodulated ONUs and a Coherent Receiver OLT," in 37th European Conference and Exposition on Optical Communications, paper Tu.5.C.1 (2011).

Acknowledgments

At this point, I would like to acknowledge the valuable support provided by others in the last years and during the preparation of this thesis.

First of all, I want to thank my supervisors Prof. Jürg Leuthold and Prof. Wolfgang Freude for providing vision, guidance and funding for my work. Many thanks go out to Prof. Leuthold for his support in every respect, many valuable discussions, and his open mind for new ideas. Finally I would like to thank him for giving me the opportunity to perform high level research using some of the best equipment available. Prof. Freude, I would like to thank him for the numerous hours we spent discussing all the published manuscripts. It is his thoroughness and mathematical skills that significantly contributed to the quality of the papers. Furthermore, the fact that we often had different points of views on a certain topic not only led to interesting discussions but also brought further insights and follow-up research work.

I want to thank Prof. Michael Hübner for introducing me to the world of digital signal processing with FPGAs. I thank for his support during my first steps towards VHDL-programming and for providing the latest state-of-the-art FPGAs to work with.

I would like to thank Dr. Bernd Nebendahl and Michael Königsmann from Agilent Technologies for 4 years of great collaboration. I enjoyed the exhibitions we put together at ECOC and OFC over the years as well as the time I measured for the first OFC paper at Böblingen, while our laboratory was under construction.

I thank Dr. Shalva Ben-Ezra for his unconventional help whenever we were short of equipment during experiments. I further thank him for the great conversations and discussions we had over the last years.

I would like to thank Dr. Matthias Frei, Dr. Tobias Ellermeyer, Lars Altenhain, and Joachim Lutz from Micram who always supported us with hardware and many practical hints while we were first setting up the transmitter and throughout the last years.

I extend my thanks to the mechanical workshop, Hans Bürger, Manfred Hirsch, Werner Höhne, and the apprentices for their great support during the construction of the software-defined optical transmitter housing. I also thank Martin Winkeler who significantly contributed to the design of the transmitter chassis and electronics. It was for their support that we were able to present our optical transmitter to the optical communications community at conferences all around Europe and the United States.

My former colleagues taught me a lot that I am grateful for. I thank David Hillerkuss for setting up the first software-defined transmitter together. He introduced me into the world of optical systems and high frequency electronics. I thank Dr. Marcus Winter for numerous discussions and his support with simulations. I want to thank Dr. René Bonk for giving me insight into the field of access networks and semiconductor optical amplifiers.

I want to thank all my colleagues at IPQ for the great collaboration. I thank my teammate Philipp Schindler for the great times in both the laboratory and the office. Thanks for the support throughout numerous experiments and for the many interesting discussions. I also thank you for always helping out without hesitation, especially with respect to the Accordance project. I would like to further thank my colleagues from the ITIV, Michael Dreschmann and Joachim Meyer, for their unconventional support in all FPGA related issues as well as your great work in the Accordance and Condor projects. I thank Swen König for sharing his knowledge on RF electronics and access networks with me. I also thank Stefan Wolf for his great work in the laboratory from optimizing the loop to implementing custom algorithms in Matlab. Further thanks go out to Matthias Meyer for his great work making the real-time Nyquist pulse transmitter come true.

Many thanks to all my students who have contributed to our work: Matthias Meyer, Benedikt Bäuerle, Philipp Kleinow, Nehemia Girma, Felix Frei, and Arne Josten.

Many thanks to our secretaries, Bernadette Lehmann, Andrea Riemensperger, and Angelika Olbrich, for their caring personalities and the help in administrative matters.

I would like to thank my parents Marina and Hans-Joachim Schmogrow who always loved and supported me no matter what goal I was heading for. Without you, this work would not have been possible.

Last but not least, I would like to thank my true love Bianca Hatwieger for her unceasing love and support over all these years. Thank you for always showing sympathy and understanding for the many hours I spent on research.

This work was supported by the European projects ACCORDANCE, OTONES, EuroFOS, and FOX-C, the BMBF project CONDOR, the Xilinx University Program (XUP), OFS, Micram, Finisar, and the Agilent University Relations Program. Further support is acknowledged from the Karlsruhe School of Optics & Photonics (KSOP), and from the German Research Foundation (DFG).

List of Publications

Journal Papers

[J1] **R. Schmogrow**, M. Meyer, P.C. Schindler, M. Dreschmann, J. Meyer, S. Ben-Ezra, J. Becker, C. Koos, W. Freude, and J. Leuthold, "Real-time Nyquist signaling with dynamic precision and flexible non-integer oversampling," Opt. Express, vol. 22, no. 1 , pp. 193–209, January 2014, doi: 10.1364/OE.22.000193.

[J2] **R. Schmogrow**, B. Nebendahl, A. Josten, P.C. Schindler, C. Koos, W. Freude, and J. Leuthold, "Timing, carrier frequency and phase recovery for OFDM and Nyquist signals using a mean modulus algorithm," Opt. Express, vol. 22, no. 8 , pp. 9344–9359, April 2014, doi: 10.1364/OE.22.009344.

[J3] **R. Schmogrow**, P.C. Schindler, C. Koos, W. Freude, and J. Leuthold, "Blind Polarization Demultiplexing with Low Computational Complexity," IEEE Photon. Technol. Lett., vol. 25, no. 13, pp. 1230–1233, July 2014, doi: 10.1109/LPT.2013.2261289.

[J4] P.C. Schindler, **R. Schmogrow**, S. Wolf, B. Baeuerle, B. Nebendahl, C. Koos, W. Freude, and J. Leuthold, "Full flex-grid asynchronous multiplexing demonstrated with Nyquist pulse-shaping," Opt. Express, vol. 22, no. 9 , pp. 10923–10937, May 2014, doi: 10.1364/OE.22.010923.

[J5] P.C. Schindler, D. Korn, C. Stamatiadis, M.F. O'Keefe, L. Stampoulidis, **R. Schmogrow**, P. Zakynthinos, R. Palmer, N. Cameron, Y. Zhou, R.G. Walker, E. Kehayas, S. Ben-Ezra, I. Tomkos, L. Zimmermann, K. Petermann, W. Freude, C. Koos, and J. Leuthold, "Monolithic GaAs electro-optic IQ modulator demonstrated at 150 Gbit/s with 64QAM," J. Lightw. Technol., vol. 32, no. 4, pp. 760–765, February 2014, doi: 10.1109/JLT.2013.2278381.

[J6] J. Pfeifle, V. Brasch, M. Lauermann, Y. Yu, D. Wegner, T. Herr, K. Hartinger, P. Schindler, J. Li, D. Hillerkuss, **R. Schmogrow**, C. Weimann, R. Holzwarth, W. Freude, J. Leuthold, T. Kippenberg, and C. Koos, "Coherent terabit communications with microresonator Kerr frequency combs," Nat. Photonics, vol. 8, pp. 375–380, April 2014, doi: 10.1038/nphoton.2014.57.

[J7] C. Weimann, P.C. Schindler, R. Palmer, S. Wolf, D. Bekele, D. Korn, J. Pfeifle, S. Koeber, **R. Schmogrow**, L. Alloatti, D. Elder, H. Yu, W. Bogaerts, L.R. Dalton, W. Freude, J. Leuthold, and C. Koos, "Silicon-organic hybrid (SOH) frequency comb sources for terabit/s data transmission," Opt. Express, vol. 22, no. 3 , pp. 3629–3637, February 2014, doi: 10.1364/OE.22.003629.

[J8] A. Agmon, M. Nazarathy, D. Marom, S. Ben-Ezra, A. Tolmachev, R. Killey, P. Bayvel, L. Meder, M. Hübner, W. Meredith, G. Vickers, P. Schindler, **R. Schmogrow**, D. Hillerkuss, W. Freude, C. Koos, and J. Leuthold, "OFDM/WDM PON With Laserless, Colorless 1 Gb/s ONUs Based on Si-PIC and Slow IC," J. Opt. Commun. Netw., vol. 6, no. 3, pp. 225–237, 2014, doi: 10.1364/JOCN.6.000225.

[J9] D. Korn, R. Palmer, H. Yu, P. Schindler, L. Alloatti, M. Baier, **R. Schmogrow**, W. Bogaerts, S. Selvaraja, G. Lepage, M. Pantouvaki, J. Wouters, P. Verheyen, J. Van Campenhout, B. Chen, R. Baets, P. Absil, R. Dinu, C. Koos, W. Freude, and J. Leuthold, "Silicon-organic hybrid (SOH) IQ modulator using the linear electro-optic effect for transmitting 16QAM at 112 Gbit/s," Opt. Express, vol. 21, no. 11 , pp. 13219–13227, May 2013, doi: 10.1364/OE.21.013219.

[J10] S. Koenig, D. Lopez-Diaz, J. Antes, F. Boes, R. Henneberger, A. Leuther, A. Tessmann, **R. Schmogrow**, D. Hillerkuss, R. Palmer, T. Zwick, C. Koos, W. Freude, O. Ambacher, J. Leuthold, and I. Kallfass, " Wireless sub-THz communication system with high data rate," Nat. Photonics, vol. 7, pp. 977–981, October 2013, doi: 10.1038/nphoton.2013.275.

[J11] R. Palmer, L. Alloatti, D. Korn, P.C. Schindler, **R. Schmogrow**, W. Heni, S. Koenig, J. Bolten, T. Wahlbrink, M. Waldow, H. Yu, W. Bogaerts, R. Dinu, W. Freude, C. Koos, and J. Leuthold, "Silicon-Organic-Hybrid MZI Modulator Generating 40 Gbit/s OOK, BPSK and 84 Gbit/s 8-ASK Signals," IEEE Photon. J., vol. 5, no. 2, article no. 6600907, April 2013, doi: 10.1109/JPHOT.2013.2258142.

[J12] **R. Schmogrow**, S. Ben-Ezra, P.C. Schindler, B. Nebendahl, C. Koos, W. Freude, and J. Leuthold, "Pulse-Shaping with Digital, Electrical, and Optical Filters – A Comparison," J. Lightw. Technol., vol. 31, no. 15, pp. 2570–2577, August 2013, doi: 10.1109/JLT.2013.2271513.

[J13] S. Koenig, J. Antes, D. Lopez-Diaz, **R. Schmogrow**, T. Zwick, C. Koos, W. Freude, J. Leuthold, and I. Kallfass, "20 Gbit/s Wireless Bridge at 220 GHz Connecting Two Fiber-Optic Links," J. Opt. Commun. Netw., vol. 6, no. 1, pp. 54–61, 2014, doi: 10.1364/JOCN.6.000054.

[J14] A.K. Mishra, **R. Schmogrow**, I. Tomkos, D. Hillerkuss, C. Koos, W. Freude, J. Leuthold, "Flexible RF-Based Comb Generator," IEEE Photon. Technol. Lett., vol. 25, no. 7, pp. 701–704, April 2013, doi: 10.1109/LPT.2013.2249509.

[J15] **R. Schmogrow**, M. Winter, D. Hillerkuss, B. Nebendahl, S. Ben-Ezra, J. Meyer, M. Dreschmann, M. Huebner, J. Becker, C. Koos, W. Freude, and J. Leuthold, "Real-time OFDM transmitter beyond 100 Gbit/s," Opt. Express, vol. 19, no. 13 , pp. 12740–12749, June 2011, doi: 10.1364/OE.19.012740.

[J16] **R. Schmogrow**, R. Bouziane, M. Meyer, P. Milder, P. Schindler, R. Killey, P. Bayvel, C. Koos, W. Freude, and J. Leuthold, "Real-time OFDM or Nyquist pulse generation – which performs better with limited resources?," Opt. Express, vol. 20, no. 26, pp. B543–B551, December 2012, doi: 10.1364/OE.20.00B543.

[J17] **R. Schmogrow**, D. Hillerkuss, S. Wolf, B. Bäuerle, M. Winter, P. Kleinow, B. Nebendahl, T. Dippon, P. Schindler, C. Koos, W. Freude, and J. Leuthold, "512QAM Nyquist sinc-pulse transmission at 54 Gbit/s in an optical bandwidth of 3 GHz," Opt. Express, vol. 20, no. 6, pp. 6439–6447, March 2012, doi: 10.1364/OE.20.006439.

[J18] N. Lindenmann, G. Balthasar, D. Hillerkuss, **R. Schmogrow**, M. Jordan, J. Leuthold, W. Freude, and C. Koos, "Photonic wire bonding: a novel concept for chip-scale interconnects," Opt. Express, vol. 20, no. 16, pp. 17667–17677, July 2012, doi: 10.1364/OE.20.017667.

[J19] R. Bonk, G. Huber, T. Vallaitis, S. Koenig, **R. Schmogrow**, D. Hillerkuss, R. Brenot, F. Lelarge, G. Duan, S. Sygletos, C. Koos, W. Freude, and J. Leuthold, "Linear semiconductor optical amplifiers for amplification of advanced modulation formats," Opt. Express, vol. 20, no. 9, pp. 9657–9672, April 2012, doi: 10.1364/OE.20.009657.

[J20] R. Bouziane, **R. Schmogrow**, D. Hillerkuss, P. Milder, C. Koos, W. Freude, J. Leuthold, P. Bayvel, and R. Killey, "Generation and transmission of 85.4 Gb/s real-time 16QAM coherent optical OFDM signals over 400 km SSMF with preamble-less reception," Opt. Express, vol. 20, no. 19, pp. 21612–21617, September 2012, doi: 10.1364/OE.20.021612.

[J21] J. Li, **R. Schmogrow,** D. Hillerkuss, P. Schindler, M. Nazarathy, C. Schmidt-Langhorst, S. Ben-Ezra, I. Tselniker, C. Koos, W. Freude, and J. Leuthold, "A self-coherent receiver for detection of PolMUX coherent signals," Opt. Express, vol. 20, no. 19, pp. 21413–21433, September 2012, doi: 10.1364/OE.20.021413.

[J22] **R. Schmogrow**, M. Winter, M. Meyer, D. Hillerkuss, S. Wolf, B. Baeuerle, A. Ludwig, B. Nebendahl, S. Ben-Ezra, J. Meyer, M. Dreschmann, M. Huebner, J. Becker, C. Koos, W. Freude, and J. Leuthold, "Real-time Nyquist pulse generation beyond 100 Gbit/s and its relation to OFDM," Opt. Express, vol. 20, no. 1, pp. 317–337, December 2011, doi: 10.1364/OE.20.000317.

[J23] D. Hillerkuss, **R. Schmogrow**, M. Meyer, S. Wolf, M. Jordan, P. Kleinow, N. Lindenmann, P. Schindler, A. Melikyan, X. Yang, S. Ben-Ezra, B. Nebendahl, M. Dreschmann, J. Meyer, F. Parmigiani, P. Petropoulos, B. Resan, A. Oehler, K. Weingarten, L. Altenhain, T. Ellermeyer, M. Moeller, M. Huebner, J. Becker, C. Koos, W. Freude, and J. Leuthold, "Single-Laser 32.5 Tbit/s Nyquist WDM Transmission," J. Opt. Commun. Netw., vol. 4, no. 10, pp. 715–723, March 2012, doi: 10.1364/JOCN.4.000715.

[J24] D. Hillerkuss, **R. Schmogrow**, T. Schellinger, M. Jordan, M. Winter, G. Huber, T. Vallaitis, R. Bonk, P. Kleinow, F. Frey, M. Roeger, S. Koenig, A. Ludwig, A. Marculescu, J. Li, M. Hoh, M. Dreschmann, J. Meyer, S. Ben Ezra, N. Narkiss, B. Nebendahl, F. Parmigiani, P. Petropoulos, B. Resan, A. Oehler, K. Weingarten, T. Ellermeyer, J. Lutz, M. Moeller, M. Huebner, J. Becker, C. Koos, W. Freude, and J. Leuthold, "26 Tbit s^{-1} line-rate super-channel transmission utilizing all-optical fast Fourier transform processing," Nat. Photonics, vol. 5, pp. 364–371, June 2011, doi: 10.1038/nphoton.2011.74.

[J25] **R. Schmogrow**, D. Hillerkuss, M. Dreschmann, M. Huebner, M. Winter, J. Meyer, B. Nebendahl, C. Koos, J. Becker, W. Freude, and J. Leuthold, "Real-Time Software-Defined Multiformat Transmitter Generating 64QAM at 28 GBd," IEEE Photon. Technol. Lett., vol. 22, no. 21, pp. 1601–1603, November 2010, doi: 10.1109/LPT.2010.2073698.

[J26] **R. Schmogrow**, B. Nebendahl, M. Winter, A. Josten, D. Hillerkuss, S. Koenig, J. Meyer, M. Dreschmann, M. Huebner, C. Koos, J. Becker, W. Freude, and J. Leuthold, "Error vector magnitude as a performance measure for advanced modulation formats," IEEE Photon. Technol. Lett., vol. 24, no. 1 , pp. 61–63, January 2012, doi: 10.1109/LPT.2011.2172405.

[J27] **R. Schmogrow**, B. Nebendahl, M. Winter, A. Josten, D. Hillerkuss, S. Koenig, J. Meyer, M. Dreschmann, M. Huebner, C. Koos, J. Becker, W. Freude, and J. Leuthold, "Corrections to "Error Vector Magnitude as a Performance Measure for Advanced Modulation Formats" [Jan 1, 2012 61-63]," IEEE Photon. Technol. Lett., vol. 24, no. 23 , pp. 2198, December 2012, doi: 10.1109/LPT.2012.2219471.

Conference Contributions

[C1] **R. Schmogrow**, P. Schindler, W. Freude, and J. Leuthold, "Digital Pulse-Shaping for Spectrally Efficient and Flexible Coherent Optical Networks," in in OSA Topical Meeting: Photonic Networks and Devices (NETWORKS), San Diego, CA, USA, July 2014, paper NM4D.3 (invited).

[C2] M. Dreschmann, J. Meyer, P. Schindler, **R. Schmogrow**, J. Leuthold, W. Freude, and J. Becker, "An ultra-high speed OFDMA system for optical access networks," in Proc. International Conference on Computing, Networking and Communications (ICNC), Honolulu, HI, USA, February 2014, pp. 889–894.

[C3] C. Weimann, P. Schindler, D. Bekele, R. Palmer, D. Korn, J. Pfeifle, S. Koeber, **R. Schmogrow**, L. Alloatti, D. Elder, H. Yu, W. Bogaerts, L. Dalton, W. Freude, J. Leuthold, and C. Koos, "Data Transmission at Terabit/s Data Rates Using Silicon-Organic Hybrid (SOH) Frequency Combs," in Proc. Optical Fiber Communication Conference (OFC'14), San Francisco, CA, USA, March 2014, paper Th4I.2.

[C4] **R. Schmogrow**, P.C. Schindler, C. Koos, W. Freude, and J. Leuthold, "Pulse-shaping for spectrally efficient coherent optical networks: OFDM, Nyquist signaling, and DFT-spread OFDM," in Proc. SPIE9009, Next-Generation Optical Communication: Components, Sub-Systems, and Systems III, San Francisco, CA, USA, December 2013, doi: 10.1117/12.2044388.

[C5] **R. Schmogrow**, M. Meyer, P. Schindler, A. Josten, S. Ben-Ezra, C. Koos, W. Freude, and J. Leuthold, "252 Gbit/s Real-Time Nyquist Pulse Generation by Reducing the Oversampling Factor to 1.33," in Proc. Optical Fiber Communication Conference (OFC'13), Anaheim, CA, USA, March 2013, paper OTu2I.1.

[C6] D. Korn, P. C. Schindler, C. Stamatiadis, M. F. O'Keefe, L. Stampoulidis, **R. M. Schmogrow**, P. Zakynthinos, R. Palmer, N. Cameron, Y. Zhou, R. G. Walker, E. Kehayas, I. Tomkos, L. Zimmermann, K. Petermann, W. Freude, C. Koos, and J. Leuthold, "First Monolithic GaAs IQ Electro-optic Modulator, Demonstrated at 150 Gbit/s with 64-QAM," in Proc. Optical Fiber Communication Conference (OFC'13), Anaheim, CA, USA, March 2013, paper PDP5C.4.

[C7] S. Koenig, F. Boes, D. Lopez-Diaz, J. Antes, R. Henneberger, **R. M. Schmogrow**, D. Hillerkuss, R. Palmer, T. Zwick, C. Koos, W. Freude, O. Ambacher, I. Kallfass, and J. Leuthold, "100 Gbit/s Wireless Link with mm-Wave Photonics," in Proc. Optical Fiber Communication Conference (OFC'13), Anaheim, CA, USA, March 2013, paper PDP5B.4.

[C8] P. Schindler, **R. Schmogrow**, M. Dreschmann, J. Meyer, D. Hillerkuss, I. Tomkos, J. Prat, H.-G. Krimmel, T. Pfeiffer, P. Kourtessis, J. Becker, C. Koos, W. Freude, and J. Leuthold, "Flexible WDM-PON with Nyquist-FDM and 31.25 Gbit/s per Wavelength Channel Using Colorless, Low-Speed ONUs," in Proc. Optical Fiber Communication Conference (OFC'13), Anaheim, CA, USA, March 2013, paper OW1A.5.

[C9] S. Koenig, R. Bonk, **R. Schmogrow**, A. Josten, D. Karnick, H. Schmuck, W. Poehlmann, T. Pfeiffer, C. Koos, W. Freude, and J. Leuthold, "Cascade of 4 SOAs with 448 Gbit/s (224 Gbit/s) Dual Channel Dual Polarization 16QAM (QPSK) for High-Capacity Business Paths in Converged Metro-Access Networks," in Proc. Optical Fiber Communication Conference (OFC'13), Anaheim, CA, USA, March 2013, paper OTh4A.3.

[C10] R. Palmer, L. Alloatti, D. Korn, P. Schindler, **R. Schmogrow**, M. Baier, S. Koenig, D. Hillerkuss, J. Bolten, T. Wahlbrink, M. Waldow, R. Dinu, W. Freude, C. Koos, and J. Leuthold, "Silicon-Organic Hybrid (SOH) Modulator Generating up to 84 Gbit/s BPSK and M-ASK Signals," in Proc. Optical Fiber Communication Conference (OFC'13), Anaheim, CA, USA, March 2013, paper OW4J.6.

[C11] J. Pfeifle, M. Lauermann, D. Wegner, J. Li, K. Hartinger, V. Brasch, T. Herr, D. Hillerkuss, **R. Schmogrow**, T. Schimmel, R. Holzwarth, T. Kippenberg, J. Leuthold, W. Freude, and C. Koos, "Microresonator-Based Frequency Comb Generator as Optical Source for Coherent WDM Transmission," in Proc. Optical Fiber Communication Conference (OFC'13), Anaheim, CA, USA, March 2013, paper OW3C.2.

[C12] J. Leuthold, W. Freude, C. Koos, **R. Schmogrow**, and D. Hillerkuss, "OFDM and Multiplexing: Tbit/s Capacities and Spectral Efficiencies up to 18 bit/s/Hz," in OSA Topical Meeting: Signal Processing in Photonic Communications (SPPCom), Colorado Springs, CO, USA, June 2012, invited paper SpTu1A.4.

[C13] S. Mumtaz, G. Rekaya-Ben Othman, Y. Jaouen, J. Li, S. Koenig, **R. Schmogrow**, and J. Leuthold, "Alamouti Code against PDL in Polarization Multiplexed Systems," in OSA Topical Meeting: Signal Processing in Photonic Communications (SPPCom), Toronto, Canada, June 2011, paper SPTuA2.

[C14] B. Nebendahl, **R. Schmogrow**, T. Dennis, A. Josten, D. Hillerkuss, S. Koenig, J. Meyer, M. Dreschmann, M. Winter, M. Huebener, W. Freude, C. Koos, and J. Leuthold, "Quality Metrics in Optical Modulation Analysis: EVM and its relation to Q-factor, OSNR, and BER," in Proc. Asia Communications and Photonics Conference (ACP'12), Guangzhou, China, November 2012, invited paper AF3G.2.

[C15] S. Mumtaz, J. Li, S. Koenig, Y. Jaouen, **R. Schmogrow**, G. Rekaya-Ben Othman, and J. Leuthold, "Experimental Demonstration of PDL Mitigation using Polarization-Time Coding in PDM-OFDM Systems," in OSA Topical Meeting: Signal Processing in Photonic Communications (SPPCom), Toronto, Canada, June 2011, paper SPWB6.

[C16] **R. Schmogrow**, B. Baeuerle, D. Hillerkuss, B. Nebendahl, C. Koos, W. Freude, and J. Leuthold, "Raised-Cosine OFDM for Enhanced Out-of-Band Suppression at Low Subcarrier Counts," in OSA Topical Meeting: Signal Processing in Photonic Communications (SPPCom), Colorado Springs, CO, USA, June 2012, paper SpTu2A.2.

[C17] J. Leuthold, W. Freude, C. Koos, R. Bonk, S. Koenig, D. Hillerkuss, and **R. Schmogrow**, "Semiconductor Optical Amplifiers in Extended Reach PONs," in OSA Topical Meeting: Access Networks and In-house Communications (ANIC), Toronto, Canada, June 2011, invited paper ATuA1.

[C18] **R. Schmogrow**, M. Meyer, S. Wolf, B. Nebendahl, D. Hillerkuss, B. Baeuerle, M. Dreschmann, J. Meyer, M. Huebner, J. Becker, C. Koos, W. Freude, and J. Leuthold, "150 Gbit/s Real-Time Nyquist Pulse Transmission Over 150 km SSMF Enhanced by DSP with Dynamic Precision," in Proc. Optical Fiber Communication Conference (OFC'12), Los Angeles, CA, USA, March 2012, paper OM2A.6.

[C19] **R. Schmogrow**, R. Bouziane, D. Hillerkuss, P. Milder, R. Koutsoyannis, R. Killey, Y. Benlachtar, P. Watts, P. Bayvel, C. Koos, W. Freude, and J. Leuthold, "85.4 Gbit/s Real-Time OFDM Signal Generation with Transmission over 400 km and Preamble-less Reception," in Proc. Conf. on Lasers and Electro-Optics (CLEO'12), San Jose, CA, USA, May 2012, paper CTh1H.4.

[C20] **R. Schmogrow**, S. Wolf, B. Baeuerle, D. Hillerkuss, B. Nebendahl, C. Koos, W. Freude, and J. Leuthold, "Nyquist Frequency Division Multiplexing for Optical Communications," in Proc. Conf. on Lasers and Electro-Optics (CLEO'12), San Jose, CA, USA, May 2012, paper CTh1H.2 (invited).

[C21] **R. Schmogrow**, P. Schindler, D. Hillerkuss, C. Koos, W. Freude, and J. Leuthold, "Uplink Solutions for Future Access Networks," in OSA Topical Meeting: Access Networks and In-house Communications (ANIC), Colorado Springs, CO, USA, June 2012, paper AW4A.1 (invited).

[C22] J. Leuthold, W. Freude, C. Koos, D. Hillerkuss, **R. Schmogrow**, and S. Ben-Ezra, "Terabit/s Super-Channels Based on OFDM," in OSA Topical Meeting: Signal Processing in Photonic Communications (SPPCom), Toronto, Canada, June 2011, invited paper SPMB1.

[C23] J. Li, C. Schmidt-Langhorst, **R. Schmogrow**, D. Hillerkuss, M. Lauermann, M. Winter, K. Worms, C. Schubert, C. Koos, W. Freude, and J. Leuthold, "Self-Coherent Receiver for PolMUX Coherent Signals," in Proc. Optical Fiber Communication Conference (OFC'11), Los Angeles, CA, USA, March 2011, paper OWV5.

[C24] P. Schindler, **R. Schmogrow**, D. Hillerkuss, M. Nazarathy, S. Ben-Ezra, C. Koos, W. Freude, and J. Leuthold, "Remote Heterodyne Reception of OFDM-QPSK as Downlink-Solution for Future Access Networks," in OSA Topical Meeting: Access Networks and In-house Communications (ANIC), Colorado Springs, CO, USA, June 2012, paper AW4A.3.

[C25] D. Hillerkuss, **R. Schmogrow**, M. Huebner, M. Winter, B. Nebendahl, J. Becker, W. Freude, and J. Leuthold, "Software-Defined Multi-Format Transmitter with Real-Time Signal Processing for up to 160 Gbit/s," in OSA Topical Meeting: Signal Processing in Photonic Communications (SPPCom), Karlsruhe, Germany, June 2010, paper SPTuC4.

[C26] **R. Schmogrow**, R. Bouziane, M. Meyer, P. Milder, P. Schindler, P. Bayvel, R. Killey, W. Freude, and J. Leuthold, "Real-Time Digital Nyquist-WDM and OFDM Signal Generation: Spectral Efficiency Versus DSP Complexity," in Proc. European Conference and Exhibition on Optical Communication (ECOC'12), Amsterdam, Netherlands, September 2012, paper Mo.2.A.4.

[C27] **R. Schmogrow**, M. Winter, B. Nebendahl, D. Hillerkuss, J. Meyer, M. Dreschmann, M. Huebner, J. Becker, C. Koos, W. Freude, and J. Leuthold, "101.5 Gbit/s Real-Time OFDM Transmitter with 16QAM Modulated Subcarriers," in Proc. Optical Fiber Communication Conference (OFC'11), Los Angeles, CA, USA, March 2011, paper OWE5.

[C28] N. Lindenmann, G. Balthasar, M. Jordan, D. Hillerkuss, **R. Schmogrow**, W. Freude, J. Leuthold, and C. Koos, "Low Loss Photonic Wire Bond Interconnects Enabling 5 TBit/s Data Transmission," in Proc. Optical Fiber Communication Conference (OFC'12), Los Angeles, CA, USA, March 2012, paper OW4I.4.

[C29] R. Bonk, G. Huber, T. Vallaitis, **R. Schmogrow**, D. Hillerkuss, C. Koos, W. Freude, and J. Leuthold, "Impact of alfa-factor on SOA Dynamic Range for 20 GBd BPSK, QPSK and 16-QAM Signals," in Proc. Optical Fiber Communication Conference (OFC'11), Los Angeles, CA, USA, March 2011, paper OML4.

[C30] **R. Schmogrow**, M. Winter, M. Meyer, D. Hillerkuss, B. Nebendahl, J. Meyer, M. Dreschmann, M. Huebner, J. Becker, C. Koos, W. Freude, and J. Leuthold, "Real-Time Nyquist Pulse Modulation Transmitter Generating Rectangular Shaped Spectra of 112 Gbit/s 16QAM Signals," in OSA Topical Meeting: Signal Processing in Photonic Communications (SPPCom), Toronto, Canada, June 2011, paper SPMA5.

[C31] D. Hillerkuss, T. Schellinger, **R. Schmogrow**, M. Winter, T. Vallaitis, R. Bonk, A. Marculescu, J. Li, M. Dreschmann, J. Meyer, S. Ben-Ezra, N. Narkiss, B. Nebendahl, F. Parmigiani, P. Petropoulos, B. Resan, K. Weingarten, T. Ellermeyer, J. Lutz, M. Moeller, M. Huebner, J. Becker, C. Koos, W. Freude, and J. Leuthold, "Single Source Optical OFDM Transmitter and Optical FFT Receiver Demonstrated at Line Rates of 5.4 and 10.8 Tbit/s," in Proc. Optical Fiber Communication Conference (OFC'10), San Diego, CA, USA, March 2010, post-deadline paper PDPC1.

[C32] J. Leuthold, M. Winter, W. Freude, C. Koos, D. Hillerkuss, T. Schellinger, **R. Schmogrow**, T. Vallaitis, R. Bonk, A. Marculescu, J. Li, M. Dreschmann, J. Meyer, M. Huebner, J. Becker, S. Ben-Ezra, N. Narkiss, B. Nebendahl, F. Parmigiani, P. Petropoulos, B. Resan, A. Oehler, K. Weingarten, T. Ellermeyer, J. Lutz, and M. Möller, "All-Optical FTT Signal Processing of a 10.8 Tb/s Single Channel OFDM Signal," in Proc. Photonics in Switching, Monterey, CA, USA, July 2010, invited paper PWC1.

[C33] W. Freude, D. Hillerkuss, T. Schellinger, **R. Schmogrow**, M. Winter, T. Vallaitis, R. Bonk, A. Marculescu, J. Li, M. Dreschmann, J. Meyer, S. Ben-Ezra, M. Caspi, B. Nebendahl, F. Parmigiani, P. Petropoulos, B. Resan, A. Oehler, K. Weingarten, T. Ellermeyer, J. Lutz, M. Moeller, M. Huebner, J. Becker, C. Koos, and J. Leuthold, "All-optical Real-time OFDM Transmitter and Receiver," in Proc. Conf. on Lasers and Electro-Optics (CLEO'11), Baltimore, MD, USA, May 2011, invited paper CThO1.

[C34] M. Dreschmann, J. Meyer, M. Huebner, **R. Schmogrow**, D. Hillerkuss, J. Becker, J. Leuthold, and W. Freude, "Implementation of an ultra-high speed 256-point FFT for Xilinx Virtex-6 devices," in Proc. 9th IEEE International Conference on Industrial Informatics (INDIN'11), Lisbon, Portugal, July 2011, pp. 829–834.

[C35] W. Freude, **R. Schmogrow**, B. Nebendahl, M. Winter, A. Josten, D. Hillerkuss, S. Koenig, J. Meyer, M. Dreschmann, M. Huebner, C. Koos, J. Becker, and J. Leuthold, "Quality metrics for optical signals: Eye diagram, Q-factor, OSNR, EVM and BER," in 14th International Conference on Transparent Optical Networks (ICTON'12), Coventry, UK, July 2012, doi: 10.1109/ICTON.2012.6254380.

[C36] W. Freude, **R. Schmogrow**, D. Hillerkuss, J. Meyer, M. Dreschmann, B. Nebendahl, M. Huebner, J. Becker, C. Koos, and J. Leuthold "Reconfigurable optical transmitters and receivers," in Proc. SPIE Photonics West (OPTO-SPIE'12), San Francisco, CA, USA, February 2013, invited paper 8284-9.

[C37] C. Koos, J. Leuthold, W. Freude, T. J. Kippenberg, J. Pfeifle, C. Weimann, T. Herr, K. Hartinger, R. Holzwarth, D. Hillerkuss, **R. Schmogrow**, "Terabit/s data transmission using frequency combs," in Proc. SPIE Photonics West (LASE-SPIE'13), San Francisco, CA, USA, February 2013, invited paper 8629-24.

[C38] C. Koos, J. Leuthold, W. Freude, N. Lindenmann, S. Koeber, G. Balthasar, J. Hoffmann, T. Hoose, P. Huebner, D. Hillerkuss, **R. Schmogrow**, "Photonic wire bonding: Connecting nanophotonic circuits across chip boundaries," in Proc. SPIE Photonics West (MOEMS-MEMS-SPIE'13), San Francisco, CA, USA, February 2013, invited paper 8613-31.

[C39] C. Koos., J. Leuthold, W. Freude, L. Alloatti, D. Korn, R. Palmer, M. Lauermann, N. Lindenmann, S. Koeber, J. Pfeifle, P. C. Schindler, D. Hillerkuss, **R. Schmogrow**, "Silicon-Organic Hybrid Integration and Photonic Wire Bonding: Technologies for Terabit/s Interconnects," Joint Symposium on Opto- and Microelectronic Dvices and Circuits (SODC'12), Hangzhou, China, September 2012, invited paper.

[C40] J. Leuthold, **R. Schmogrow**, D. Hillerkuss, C. Koos, W. Freude, "Super Channels Based on Nyquist Multiplexing," in Proc. 17th Opto-Electronics and Communications Conference (OECC'12), Busan, South Korea, July 2012, invited paper 3B1-2.

[C41] A. K. Mishra, Z. Wang, H. Klein, R. Bonk, S. Koenig, D. Karnick, **R. Schmogrow**, D. Hillerkuss, M. Moehrle, T. Pfeiffer, C. Koos, J. Leuthold, W. Freude, "Performance analysis of an OFDM transmission system with directly modulated lasers for wireless backhauling," in Proc. 14th Intern. Conf. on Transparent Optical Networks (ICTON'12), Coventry, UK, July 2012, doi: 10.1109/ICTON.2012.6254423.

[C42] **R. Schmogrow**, B. Nebendahl, D. Hillerkuss, C. Koos, W. Freude, and J. Leuthold, "Quality metrics for advanced modulation formats in optical communications: OSNR, Q-factor, EVM, and BER," in Proc. 13. ITG-Fachtagung, Photonische Netze, Leipzig, Germany, May 2012.

Resume

René Marcel Schmogrow
rms.09.83@gmail.com

Work Experience

09/2013–present	**Infinera Corporation, Sunnyvale, CA, USA** Sr. HW Development Engineer - Development of next generation coherent optical transceivers.
05/2013–07/2013	**ETH Zürich, Institute of Electromagnetic Fields (IEF), Zurich, Switzerland** Post-doctoral research - Development of FPGA simulation platform for real-time DSP - Implementation of a Matlab simulation environment for coherent optical communications - RTL design and implementation for real-time digital filters needed to receive Nyquist signals.
11/2009–05/2013	**Karlsruhe Institute of Technology (KIT), Institute of Photonics and Quantum Electronics (IPQ), Karlsruhe, Germany** PhD studies - Development of algorithms for real-time optical transmitters and receivers - Design and development of a software-defined optical transmitter - Experimental demonstration of real-time OFDM and Nyquist generation beyond 100 Gbit/s and several optical systems - Submission of two patents for optical receivers - Major contribution to collaborations with several industrial partners including Agilent Technologies, Micram, and Finisar. - Contribution to work packages and deliverables of several European and national research projects - Tutorial lectures on optical communications (semiconductor devices, nonlinear optics) - Supervision of student projects and master theses

10/2007–04/2009 **Robert Bosch GmbH, Bühl, Germany**
Werkstudent (intern)
- Statistical analysis of measurement accuracies to specify calibration routines for custom build measurement systems in the field of engine temperature management

10/2004–09/2007 **Robert Bosch GmbH, Bühl, Germany**
BA-student (student trainee)
- Work within the field of software system testing for windshield wiper- and rain sensor electronics, software development in Labview, C#.net and UML as well as in the field of measurement technology and statistical analysis

Education

11/2009–05/2013 **Universität Karlsruhe (TH) / Karlsruhe Institute of Technology (KIT), Institute of Photonics and Quantum Electronics (IPQ)**
Ph.D. thesis on "Real-time Digital Signal Processing for Software-defined Optical Transmitters and Receivers."
Date of defense: May 7, 2013.

10/2007–10/2009 **University of Karlsruhe (TH) / Karlsruhe Institute of Technology (KIT), Karlsruhe School of Optics and Photonics (KSOP)**
M.Sc. Optics and Photonics
Grade 1.1 (sehr gut, "very good")
Master thesis: "Implementation of an OFDM capable Multiformat Transmitter"
Grade 1.0 (sehr gut, "very good")

10/2004–09/2007 **University of Cooperative Education (BA) Karlsruhe**
Dipl.-Ing. (BA)
Grade 1.1 (sehr gut, "very good")
Diploma thesis: "Realization of an intelligent wide range current acquisition system with high-speed auto-ranging"
Grade: 1.0 (sehr gut, "very good")

07/2003–03/2004 **Military service**
Military communications (radio link systems) "Fernmeldebataillon 610" Rendsburg, Germany

09/2001–07/2003	**Hans-Carossa Oberschule (Gymnasium) Berlin-Spandau (secondary school)** Allgemeine Hochschulreife (general qualification for university entrance) Grade: 1,8 (gut, "good")
2000–2001	**Junior year at a high school in the State of Louisiana, USA**